Mathematics in Mind

The monographs and occasional textbooks published in this series tap directly into the kinds of themes, research findings, and general professional activities of the **Fields Cognitive Science Network**, which brings together mathematicians, philosophers, and cognitive scientists to explore the question of the nature of mathematics and how it is learned from various interdisciplinary angles. Themes and concepts to be explored include connections between mathematical modeling and artificial intelligence research, the historical context of any topic involving the emergence of mathematical thinking, interrelationships between mathematical discovery and cultural processes, and the connection between math cognition and symbolism, annotation, and other semiotic processes. All works are peer-reviewed to meet the highest standards of scientific literature.

Serdar Ş. Güner

Art and IR Theory

Visual Semiotic Games

 Springer

Serdar Ş. Güner
Department of International Relations
Bilkent University
Ankara, Ankara, Türkiye

ISSN 2522-5405 ISSN 2522-5413 (electronic)
Mathematics in Mind
ISBN 978-3-031-32344-7 ISBN 978-3-031-32342-3 (eBook)
https://doi.org/10.1007/978-3-031-32342-3

Mathematics Subject Classification: 91A05, 91A80, 91A40, 91A28

This Springer imprint is published by the registered company Springer Nature Switzerland AG
The registered company address is: Gewerbestrasse 11, 6330 Cham, Switzerland

For Yunus Batu and İrem

Preface

This book originated in 2003 when I made a presentation to my colleagues and students on correspondence I claimed to exist between paintings and international relations (IR) theories. My argument was that paintings reflect esthetic and artistic approaches paralleling main ideas of IR theories. Some paintings were argued to stand for central theoretical ideas taking abstraction, color theory, and painting styles as axes of interpretation. For example, I argued that Jackson Pollock and Mark Rothko paintings represent constructivist and structural realist propositions by Alexander E. Wendt and Kenneth N. Waltz, respectively. René Magritte's chef-d'oeuvre "The Castle of Pyrénées" was also noted in my presentation for its duplication of the idealist proposition of the achievement of world peace through international institutions. The castle carved from the rock that levitates over a turbulent sea means a fantastic dream, a dream of an international organization prohibiting wars and allowing for a perpetual peace among states.

The presentation incorporated visual semiotic ideas I was unaware of by then. It took almost two decades for me to get acquainted with semiotics and visual semiotics. My efforts finally bore fruit in two publications in *Semiotica*, the premier journal of semiotics, in 2019 and 2021. The first paper offered Rothko–Waltz signifier-signified relation as a metaphoric sign to discuss structural constraints. The second offered Pollock–Wendt association as a sign expanding social constructivist viewpoint about anarchy. Working on visual semiotics and entering the domain of hermeneutics, I started to read and work on sources of philosophy of science that is the bedrock of both structural realism and constructivism. Game theory as the interactive decision theory pushed me in turn to work in the direction of the philosophy of mind. My readings in the philosophy of mind and games as interactions among decision makers revealed an area ripe to explore, the area of players' minds and mental pictures players or minds form. Do we have to focus on players' minds as distinct from their bodies or players' mind and brains constitute inseparable sides of the same coin in forming visual signs through games? How do players' minds or brains or both work in their strategic interactions? This were central questions to answer. Chapter 8 includes my alternative answers to these questions through interpretations of game equilibria.

I am in deep gratitude to Marcel Danesi who offered me a chance to contribute to the Mathematics in Mind series. It is impossible to describe how valuable his encouragement and help are. I thank him for letting me run freely in the fields of abstract art, semiotics, and game theory that led me to explore uncharted areas of scientific and interpretive inquiry of Structural Realism and Constructivism. I am deeply indebted to him. I am grateful to Nicholas DiBenedetto, the Associate Editor of the Mathematics in Mind Series of Springer Nature, for his reading the manuscript and commendation of it. I thank him for his support, encouragement, and help in the editing and publication process.

I received no financial assistance or a sabbatical leave in writing this manuscript. The pandemic posed as a serious hindrance for travels starting in 2020. But I cannot complain. It is a grand pleasure to be publishing with Springer Nature and writing a book that intersects IR theories, abstract art, visual semiotics, philosophy of science, and philosophy of mind. My conviction is that the book opens alternative roads of interpretation of IR theories and their development contributing to the Discipline of IR as a whole.

Ankara, Türkiye   Serdar Ş. Güner

Contents

Chapter 1
Scope and Method

This book has started with a simple question: could two international relations (IR) scholars coordinate by forming signs of theoretical propositions through abstract paintings? To answer the question, the book engages in the interpretation of central propositions of two IR theories constituting leading structuralist lenses to analyze IR, namely, Structural Realism proposed by Kenneth N. Waltz (Waltz 1979) and Social Constructivism by Alexander E. Wendt (Wendt 1992, 1999).

The proposition by Waltz offers a materialist perspective: it posits that the distribution of capabilities across states and anarchy, that is, the principle of the organization of international politics corresponding to the inexistence of a superior authority over co-existing sovereign states, generate structures of international politics that create constraints shaping and shoving states' interactions. The proposition by Wendt offers a social perspective: states' shared ideas about each other correspond to the condition of intersubjectivity that creates multiple anarchies. States, depending on how they develop ideas about each other through their interactions, can co-exist in a Hobbesian anarchy where each state is a potential enemy, in a Lockean anarchy where each state is a rival, or in a Kantian anarchy where each state is a friend. The three anarchies spawn their own constraints upon states' interactions.

The book contributes to the recent burgeoning of visual culture in the Discipline in an innovative and interpretive fashion. Its aim is not scientific prediction or falsification but comprehension (Aumann 1985, 6). It proposes that the distance between art and abstract IR ideas becomes shorter as theorists connect, associate, and relate features of abstract art with ideas of Waltz and Wendt. Aumann (1985, 6) indicates that "To 'understand' an idea or a phenomenon – or even something like a piece of music – is to relate it to familiar ideas or experiences, to fit it into a framework in which one feels 'at home.'" Borrowing Aumann's words, I presuppose that theorists make sense of their familiar theoretical contentions about Waltz and Wendt by relating them to Rothko and Pollock's art, respectively. They discover new theoretical horizons by understanding art as making sense for IR theories.

© Springer Nature Switzerland AG 2023
S. Ş. Güner, *Art and IR Theory*, Mathematics in Mind,
https://doi.org/10.1007/978-3-031-32342-3_1

Art and the Discipline are not disconnected realms. Paintings affirm that "many non-textual features of human life such as actions, tools, social roles and individual lives, can and should be taken as meaningful in the same way as texts are" (Rouse 1987, 42). Thus, paintings prove that a common theoretical understanding through art is possible in the discipline; paintings generate meanings helping to translate linguistic thoughts into plastic ones.

Waltz's book *Theory of International Politics* (*TIP*) occupies a prominent place within the IR theory literature. The book has attracted widespread criticism and acclaim over these 44 years passed since its publication. The distinction of *TIP* is well noticed by scholars: "*TIP* is undoubtedly the most important book produced in the discipline within the last 60 years, possibly ever (at least referring to the discipline in its modern form)" (Wæver 2009, 203); "A great deal has been learned about international politics since Waltz wrote his famous book, but there are still many disagreements about how to understand it" (Wagner 2007, ix). Wagner further notes that "while there is a widespread agreement that much that Waltz wrote must be revised, there is much less agreement about what revisions should be made. This fact not only *impedes intellectual progress* but also diminishes the ability of scholars in the field to speak with authority to a broader audience" (Wagner 2007, x).[1]

TIP constitutes a central source of insight in the Discipline of IR indeed (referred to as the Discipline from hereon). Wendt's criticism of Waltz as *TIP* omitting social dimensions of international politics has opened a new theoretical perspective that exposes how Waltz's thinking has shaped Wendt's construction of the world (Wendt 1992, 1999). Wendt has proposed contra Waltz that distribution of shared ideas in the sense of what states think they know about each other depending on their interactions as mattering more than distribution of power taken as the prime factor of international politics by Waltz. Dynamic nonunique modes and patterns characterize how sovereign states coexist and interact according to Wendt. Wendt's contribution to the discipline is located in the intersection of sociology and IR theories. Both Structural Realism of Waltz and Social Constructivism (Constructivism in short) of Wendt remain as the two major system-level theories constituting a fundamental debate in the Discipline.

1.1 Hybrid Method

Visual semiotic approach bolstered by game theory enriches Structural Realism-Constructivism debate in an interpretive and rigorous fashion. The book proposes not a theory but a hybrid method that combines game theory and visual semiotics. The method dwells into how abstract paintings help IR theorists to communicate through a new channel. IR scholars interact by making meanings of paintings by Pollock and Rothko for propositions of Waltz and Wendt. Covers of IR Theory

---

[1] Italics are mine.

textbooks constitute a real-time gauge of how IR theories and abstract art are mutually connected. These covers reflect visual and mental character of theoretical thinking of IR authors in the Discipline. Signs players construct communication they are engaged in enrich structuralist perspectives Structural Realism and Constructivism. The method transforms visual semiotics into a rigorous interactive method. It paves the way to interpretations of game equilibria in terms of the philosophy of mind and types of explanations.

Semiotics offers analytically rich and conceptually elaborate methods (Ball and Smith 1992) yet it is unclear how to apply it (Slater 1998). The hybrid method clarifies how to apply semiotics in interactive settings. As result, the method sheds light on the debate of Structural Realism versus Constructivism inciting new questions and answers enriching the Discipline.

1.1.1   Visual Semiotics

Bal and Bryson (1991, 73) indicate that "semiotics is antirealist." Anti-realism rejects the existence of knowable mind-independent facts, objects, or properties (Audi 2015, 40–41). Therefore, semiotics rejects the existence of knowable mind-independent facts, objects, or properties. The hybrid method shows that game theory is compatible with antirealism and semiotics.

Antirealist quality of semiotics does not contradict game theory for three principal reasons. First, game theory disciplines analyses of interactions by offering a formal and interactive language. According to Cartesian dualism, players who interact correspond to interacting minds. Second, game theory contributes to semiotics' analytical richness and conceptual elaborateness by providing a formal language that clarifies how people interactively make meanings and senses of things. Third, game theory is a methodological tool compatible with the examination of social and biological problems in contrast with physical sciences where the nature is inanimate (Nowak 2006; Gintis 2000; Maynard Smith 1982; Hofbauer et al. 1979). The examination of interactive minds making meanings belongs to the category of problems about animate nature and interactive consciousness.

Visual semiotics offers communication channels posing problems and soliciting solutions (Buyssens 1967; Eco 1976, 1984; Greimas 1976; Jacobson 1960; Lotman 1990). Yet one can hold the view that IR theories and abstract art are far distant from each other as there exist infinitely many individual and subjective interpretations in abstract art. Countless interpretations do not preclude meaningful interactive communication, however. A translation mechanism substantiated in presentations of "the unpresentable" (Lyotard 1993, 7) through visual semiotics renders sign making possible and meaningful for IR theories through paintings and games. The concepts of signifier and signified of Ferdinand de Saussure and icon, index, symbol concepts of Charles Sanders Peirce help to tame the "unrepresentable" and open doors to a theoretical and disciplined enrichment in the Discipline.

How could paintings be connected with Waltz's or Wendt's abstract propositions? Sylvester (1991, 549) offers a clue to answer the question:

> Blue Poles (1952), for example, suggests several logics of anarchy producing a type of order on the verge of anti-social disorder that is saved by constraining poles inserted at the end. The social is intimated by those agentic poles; so too it is there in pieces of beer bottles encrusted in the paint, remnants of an especially drunken evening around the canvas. Pollock also draws figures of people onto many of his empty canvases and then saturates them in paint as he composes abstractions. Those figures might seem to down and disappear (he also sliced them out of some paintings such as his 1948 Untitled (Cut-Out Figure). In fact, Pollock's methods amplify the richness of a sociality that cannot be eliminated by abstraction so much as absorbed into it and reflected back. *Wendt has that Pollockian absorption of the social in his theory but is less masterful in lighting the abstractions so that they avoid opacity.*[2]

Sylvester's clue incites semiotic thoughts about how an IR scholar creates a meaning through a painting for Wendt's conceptualization of anarchy. Semiotics is the study of signs that are meanings humans attach to objects through interpretation (Danesi 2018; Chandler 2007; Sebeok 2001; Nöth 1990). It constitutes an interpretive method (Yanow 2006a, b). An immediate question arises as what outcomes one would expect if two theorists were interacting making similar interpretations of Blue Poles. How would both scholars interact by creating meanings of Wendt's proposition by interpreting Blue Poles? What art features in Blue Poles could IR theorists associate with abstract IR concepts? Without being art experts, they might recognize color swirls and moves in Pollock's drip paintings as hinting at Wendt's concept of dynamic nonunique anarchies (Wendt 1992). In a similar fashion, they could recognize patterns of color interactions in color-pane paintings by Rothko as being reminiscent of Waltz's figure displaying structural constraints (Waltz 1979, 40).

1.1.2 Visual Semiotic Games

How to study a semiotic interaction between two theorists who interpret and recognize patterns in paintings? The answer is the use of game theory that constitutes a mathematical method to study interactive decision making (Von Neumann and Morgenstern 1953; Luce and Raiffa 1957; Aumann 1985). Being a branch of mathematics, game theory forces rigor in semiotic reasoning of making signs. It constitutes a disciplined language to model interdependent sign making based on scholars' minds, individual tastes, and subjectivities. The interaction between the theorists constitutes a game in which each theorist must ponder on what meaning the other concocts through art and each choice of meaning depends on the choice of the other, hence theorists are strategically interdependent.

---

[2] Italics are mine.

Semiotics and game theory generate together a powerful hybrid method helping to examine and explore by modeling how strategically interdependent IR theorists communicate and settle on a specific meaning of an IR theory proposition through abstract art. Osborne (2004, 3) affirms that "Game-theoretic modeling starts with an idea related to some aspect of the interaction of decision-makers. We express this idea precisely in a model, incorporating features of the situation that appear to be relevant. This step is art." Therefore, the hybrid method constitutes art.

The characteristic aspect of the hybrid method involves interacting decision makers who are IR theorists and who decide what meanings they can devise and fabricate for propositions set forth by Waltz and Wendt through Rothko's and Pollock's artwork, respectively. The method permits to study processes of semiosis over IR theory propositions through abstract art and interactive rationality rigorously. It allows to decipher signs interacting players can develop through art over propositions' linguistic meanings.

I call "visual semiotic games" (VSGs from hereon) those interactive decision-making models where players attach meanings to paintings interactively as they make sense of them: "a shared tenet of embodied, embedded, situated and distributed cognition is that people make sense of things interactively" (Clark et al. 2011, 290). Players exchange signs they form along the lines of "an elaborate analytical vocabulary" (Rose 2001, 70) developed by Ferdinand de Saussure (Saussure 1996) and Charles Sanders Peirce (Peirce 1960a, b, c). Hence, paintings become sources of signs through players' faculties of interpretation (Marin 2005, 5). Signs enrich the Discipline by displaying interdependent hermeneutics.

My aim is not to test hypotheses derived from VSGs against data. Wendt (1998, 111) succinctly remarks that rational choice models "are thought to explain insofar as they capture the properties and dispositions of the systems they represent, even if they do not relate to independently existing causes." Systems in this sense correspond to VSGs. Explanations VSGs generate correspond to Nash equilibria alternative VSG models imply. No experiments or simulations are needed to divulge and interpret equilibrium conditions. The rules of VSGs explain how players' preferences over their actions that are in turn based on their preferences over the outcomes of their interaction imply certain conditions of stability in forming signs for IR theory propositions using abstract paintings. The rules of the VSGs are the rules of the levels of game theoretic analysis that are strategic and extensive forms. They are "internal structures" of the VSGs (Wendt 1998, 112). There is also the social structure of VSGs that correspond to players' interdependent mental calculations given players' preferences over the outcomes. Social structures of conditions of stability in players' interactions generate Nash equilibrium.

A VSG is a model of a semiotic interpretive process where interdependent IR theorists are involved. It incorporates players' interpretations of a painting as if they correspond to signs of theoretical statements. Strategically interdependent IR scholars make interpretive signs for IR theory propositions using abstract art. They are conscious of how their strategy choices affect each other. No player's individual subjectivity can unilaterally determine the sign correspondence between the painting and IR theory sentence. Players are aware that they cannot seal the fate of their interaction alone; sign making is not an individual, isolated matter. No player can rule out meanings the other makes as no player can solely determine the sign of a painting. Players are conscious of their strategic interdependence. They must uncover and ponder about the subjectivities of the other player. Keohane (2000, 128) affirms that "coordination takes place as a result in part of the ideas people have not only about their own beliefs but about the beliefs of others." Consequently, the principal question revolves around the interactive meaning of a painting as connected with an IR theory proposition.

Semiotic interpretive processes result in formation of signs coming out of players' strategy choices leading to joint agreement and coordination based on players' "primary reasons" (Davidson 1963, 685) that are pairs of players' beliefs that the Discipline needs a common language and players' pro attitudes indicating the appealing aspect coordination. VSGs constitute tools to study how players' sign proposals affect the emergence of coordination and a common semiotic language inexistent at the start. Consequently, all VSGs are supposed to be coordination games, that is, games in which players have an interest in avoiding strategy combinations leading to different sign choices.

Players interact on the basis of their ability to form signs which depend on their minds. Interactive meanings correspond to plasticity corresponding to mental forms (Churchland 1979) of accepting or rejecting type of signs players propose through paintings. The equilibria of VSGs demonstrate how purely linguistic IR theory statements transform into visual semiotic statements.

The rules of VSGs are common knowledge, that is, players know the number of strategies available to players, players' preferences, and the number of players in the game; a player knows that the other knows the VSG rules, one knows that the other knows that one knows the rules, and ad infinitum. The condition of common knowledge is not alien to constructivist thinking. Wendt (2000, 179) provides a clear connection between Constructivism and game theory and therefore VSGs: a "reading rationalist treatments of culture as 'common knowledge', which looks a lot like the constructivist's 'intersubjective understandings.' A player can misperceive and be uncertain about the other player's preferences over the outcomes of the interaction in a VSG modeling incomplete information, but common knowledge still rules. Thus, if one is unsure about the other's preferences like whether the other prefers to interpret a painting as a signifier, icon, index, or a symbol for a theoretical sentence, then the other is informed of the one's uncertainty and one knows that the other knows one's uncertainty, and so on.[3]

Players are rational in the simplest sense that they have consistent preference orderings over the outcomes of their interaction. If players have preferences over alternative meanings and if these preferences are consistent, it is only natural to expect them to make choices in line with those preferences as indicated by O'Neill (2001, 261): "Another interpretation of 'are people rational?' would see it as asking whether people maximize their utilities whatever those happen to be. However,

---

[3] This is called the "Harsanyi doctrine" in game theory.

there is no such thing as having a utility function but not maximizing it. Since utilities represent choices, to have a utility function is to have a certain pattern in one's disposition to make choices, and someone who does not have such pattern does not have a utility function at all." Thus, players do not maximize their satisfaction levels emanating from the outcomes, therefore their utilities as if they were computing machines without having any feelings, perceptions, or subjectivities. Interacting players do not solve complex mathematical statements translating their preferences. Instead, players have consistent preferences over assigning meanings to abstract paintings alongside with Saussurean and Peircian semiotics. Accordingly, players are assumed to have patterns prompting them to choose some paintings as sources of precise meanings. A player, for example, might view a painting by Mark Rothko as radiating or not a clear meaning of icon for Waltz's proposition according to her transitive preferences.

I propose two classes of VSGs: Saussurean and Peircian games. Two players are modeled to interact in both classes of games expressed in strategic and extensive forms. There exist three elements of VSGs in strategic form: the number of players, strategies, and players' preferences over the outcomes of their interaction. They are simultaneous-move games because no player is informed of what strategy choice the other player has made when she selects a strategy. This condition is called as imperfect information. Each player is assumed to possess two strategies in Saussurean games in strategic form: she either accepts or rejects that a painting constitutes a signifier for a specific IR theory proposition. Each player has three strategies in Peircian games in strategic form: icon, index, and symbol. Therefore, Saussurean games in strategic form are 2×2 games and Peircian games in strategic form are 3×3 games. Saussurean and Peircian games at extensive form allow players actions of signifier and no signifier and actions of icon, index, and symbol at their information sets, respectively. They contain more elements than VSGs in strategic form. They include players' information conditions, sequences of their action under those information conditions, and the specification of when the interaction ends. They are sequential-move games permitting players to be informed of previous choices made in the game, so that players might have perfect information during the interaction. It is possible to translate VSGs in extensive form into strategic form by using actions as components of a strategy as a "complete plan that specifies all possible courses of action of a player for whatever contingencies may arise" in the interaction (Brams 1985, 158). Saussurean and Peircian games constitute an art of making signs for IR theories through abstract art. They illustrate an achievement of a balance between game theory and visual semiotics and expose a novel theoretical language and perspectives in the Discipline.

The book is organized into three parts. The first part discusses the state of the Discipline in constant turns and swirls. Chapter 2 introduces Waltz's and Wendt's theoretical conceptualizations of anarchy and structural constraints. It lays the foundation for the need to solve coordination problems among IR scholars. The third chapter on the connection between IR theories and abstract art completes the first part. The second part contains Chaps. 3, 4, 5, 6, and 7 that introduce VSG models and present possible stability conditions through equilibrium concepts ranging from

Nash equilibrium to evolutionarily stable strategies (ESS). The third part is Chap. 8 where the equilibria reached in previous chapters are interpreted according to dualism and materialism, two principal doctrines of the philosophy of mind and their alternative branches. The last chapter concludes the discussion.

Chapter 2
Structural Constraints and Nonunique Dynamic Anarchies

The research agenda of the Discipline is context-dependent and sensitive to shock-waves such as world wars, 1973 Oil Crisis, Iranian 1979 revolution, Soviet Union's dissolution, and 9/11 attacks. Unexpected global scourges push IR scholars to invent new theories or to change theories they use entirely (Güner 2012). One should expect the emergence of new IR theories or renewals of old approaches and methods following catastrophic events like Covid-19 pandemic, Ukraine-Russia conflict and ensuing global food and energy crises, and other present and future calamities.

Next to its sensitivity to outside shocks, the discipline is also said to evolve through debates between alternative IR perspectives and theories (Schmidt 2012; Kurki and Wight 2010; Wæver 1996; 2009). In very broad terms, the first debate is usually argued to have happened between realism and idealism with the former evaluating IR from materialist power angle and the latter based on the belief that international peace is not an elusive dream but reachable through international institutions and organizations. The second debate has been between traditionalists who argued for a study of IR by case-by-case approach and behaviorists who believed that generalizations in IR are possible using formal methods and computer aids becoming available in the early 1960s. There was also a debate between Structural Realism and neoliberal institutionalism later in the 1980s with the former emphasizing systems approach, structural constraints, and the latter positing the logic of one-shot Prisoner's Dilemma (PD) game which demonstrates the possibility (or the impossibility) to reach an international cooperation depending on states' expectations of future cheating costs (Axelrod and Keohane 1985; Jervis 1985; Oye 1985; Snidal 1985).

Lapid (1989), in his discussion of IR debates, notes that post-positivism opens new research horizons in the Discipline pointing to radical theoretical and methodological novelties. Bell (2009, 22), for example, notes that "visions of global order have been produced by a vast array of people in different geographic locales, working in a wealth of media. Painting, cinema, architecture, the internet, and computer games are all potential sites of investigation."

© Springer Nature Switzerland AG 2023
S. Ş. Güner, *Art and IR Theory*, Mathematics in Mind,
https://doi.org/10.1007/978-3-031-32342-3_2

The discussion of the Discipline's evolution through debates and a large array of methods and approaches cannot detect every step of change in IR theories. It is difficult to maintain that, for example, the debate between idealism and realism is now extinct. Curiously, theoretical discussions in the Discipline do not zero in Waltz's key concept of structural constraints constituting the heart of Structural Realism.

2.1   The Discipline Is a Ghetto of War

The Discipline is turbulent, theoretical and methodological debates galore. Brown (2010: 145) summarizes the Discipline's current state by a ghetto-war metaphor:

> International relations is no longer confined to its own, self-imposed ghetto. As international relations theorists break down the walls of the ghetto – helped by their friends on the other side – and stream into the metropolis of social and political thought, expecting to merge into the life of the city and help shape its future, a disconcerting sight meet their eyes. Some of the most distinguished residents of the city have fled or turned their talents to attacking their previous home in the company of strangers from other cities and other countries, while other notables have found themselves forced to defend positions thought unassailable for two and a half millennia. In short, we – one-time international relations theorists – have joined a city under siege and riven by civil war. Moreover, neutrality is no longer possible; sides must be chosen, and positions staked out, especially since "international relations" is one of the major battlegrounds chosen by the insurgents.

Sylvester (2001: 320) offers an alternative metaphor paralleling Brown's. She argues that the Discipline reflects a "camp mentality" as there is "a sprawling multiplicity of disparate and insular intellectual camps which are ever less capable or even interested in communicating with one another." Sylvester's metaphor inspires an architectural metaphor: imagine that realism is nested in an old-fashioned building dating back to the 1940s or 1950s, liberalism still lives in a house built in the 1970s, and Constructivism, a newcomer to the neighborhood, occupies a house built in the early 1990s. Buildings' entrances are guarded similar to nightclubs so that no one can enter them without invitation or permission. Building residents avoid communication with each other and if they do they scuffle and quarrel (Legro and Moravcsik 1999; Mesquita and Morrow 1999; Martin 1999; Niou and Ordeshook 1999; Powell 1999; Walt 1999a; Walt 1999b; Feaver et al. 2000). One group of realists - the structural realists who draw inspiration from Kenneth Waltz's seminal *Theory of International Politics* - has been a particular "*target for constructivist arrows*" (Copeland 2000, 187).[1] It is also possible that communication is broken among residents of the same building.

Wæver (1996, 149) adds further theory labels like "pluralism, Marxism, interdependence, world society, radicalism, globalism, and structuralism" (there are countless other "isms") indicating a complex confusion of the Discipline. Methodological variations stemming from different philosophy of science doctrines like positivism

---

[1] Italics are mine.

and post-positivism on top of these new theories and communication failures add further complications and complexities. The debate of Structural Realism versus Constructivism is also apprehended as a war field particularly: For more than a decade realism, by most accounts the dominant paradigm in international relations theory, has been under *assault* by the emerging paradigm of Constructivism. Turbulent Discipline welcomes new theories creating extra communication problems. Wendt (2000, 180) proposes a peace in the Discipline through social theory by stating that "The point of social theory should be to put existing first-order theories of international relations into a broader context, calling attention to perhaps tacit presuppositions that may create problems, and identifying questions that have been not asked. The goal should be to facilitate substantive investigations rather than discourage them, to unify our knowledge rather than Balkanize it." VSGs facilitate the reaching of the goal Wendt describes by communication among players. They incorporate elements from both rational choice field and semiotics and expose insights for structural constraints and nonunique dynamic anarchies from an anti-realist perspective helping to construct a fruitful communication in the Discipline.

New axioms of IR theories can be conceived easily. For example, realism assumes that states are unitary and rational. If one relaxes this axiom, so that states are not rational, but they are boundedly rational and coexist as sovereign units, one can obtain a new branch of realism. Wagner (2007, 12) notes "One impediment to settling issues raised by Realists and their critics is that it is not entirely clear what Realism is. There is now an embarrassment of Realisms. There is classical Realism, neoclassical Realism, Structural Realism (aka Neorealism), human nature Realism, defensive and offensive Realism, and it may be undergoing further mutations as I write." Thus, controversies and dissensions happen to exist within the old-fashioned building of Realism and in other houses as well. Alternations of one or more realist axioms such as states are not the main units of international politics or states do not co-exist as sovereign units produce new IR approaches. Biology, evolutionary biology spawn new theories of IR (Thayer 2000; Masters 1983). To wit, one could expect an eventual entry of chemistry or artificial intelligence and virtual reality into the Discipline generating perhaps a "chemical realism" or a "virtual realism."

Another source of communication problem among IR scholars is the Discipline's openness to metatheoretical controversies. A new theory can be argued as being compatible with scientific realism and constructive empiricism or Kuhnian and Popperian philosophies of science (Bloor 1971). For example, Nicholson (2000, 184) remarks that "the field of International Relations is a mass of Lakatosian Research Programmes" indicating that there is no room for surprise that IR scholars discuss the merits of progressive versus degenerative research programs in the Discipline taking Waltz's balancing proposition as the central axis of their discussion (Schweller 1997; Christensen and Snyder 1997; Vasquez 1997).

Yet the debate of the "degeneration of political realism" constitutes only the tip of the iceberg. "The discipline seems to organize itself through a constant oscillation between grand debates and periods in-between where the previous contestants meet…None of these debates lasts forever" (Wæver 1996, 175). Some theorists who are "outsiders" to positivist research programs can evaluate, for example, Waltz's theory of Structural Realism as a "hopelessly narrow and oversimplified" theory (Buzan and Little 2001, 30) and recommend research along postpositivist directions like critical security or critical geopolitics (George and Campbell 1990). Naturally, next to realism, liberalism and Constructivism too are open to axiomatic and meta-theoretical changes inspired by alternative philosophy of science doctrines as the Discipline has a "open door" policy toward new approaches. Accordingly, a wild and wide spectrum of future IR theories will produce new debates. Communication deficiencies and difficulties in the Discipline are bound to occur and expected to rise as result.

The Discipline has never reached the maturity of having a common language. One can imagine the mess astronomy would suffer if there were multiple definitions of black holes. Economists understand the same concept hearing words of inflation, investment, or indifference curves. If IR theorists talk about concepts like power or identity, they disagree about the meaning of these central terms. Haas (1953, 447–458), for example, counts eight different meanings of balance of power, a central concept in the Discipline: balance of power meaning distribution of power, equilibrium, hegemony, stability and peace, instability and war, power politics, universal law of history, and a system and guide to policy making. The problem is that not all these meanings are compatible with each other.

The multiplicity of turns testifies erratic state of the Discipline. Hence, it is not surprising to witness that the Discipline turns like a whirling dervish. Baele and Bettiza (2021) list the visual (also known as the aesthetic turn), historical, practice, new material, and queer turns. There are also turns like the linguistic turn (Neumann 2002), the practice turn (Cornut 2017), the constructivist turn (Checkel 1998), the aesthetic turn (Bleiker 2001), the visual turn (Callahan 2015), and the relational turn (Donnely 2019).[2] Each turn adds theoretical complexity and diversity of terminology, metatheoretic bases, and methodology to the Discipline. The result is the exacerbation of communication and coordination problems which can be solved by a common language.

To illustrate, the visual turn evaluates mainstream IR theories like Realism and Liberalism as traps preventing our understanding of how individual subjectivities and ideas shape international politics. It is true that traditional IR approaches are unable to investigate how images weigh in international politics. Violence and torture photographs taken at Abu Ghraib in Iraq during the U.S. in 2003 invasion were presented as mediatic events through media and offered to the public consumption (Feldman 2005; Butler 2007). The emphasis was put on the place of images in the eyes of the global public with the central idea being that Abu Ghraib photographs might attract public attention and recognition through media. They might perhaps have found some reaction in American society and American culture (Giroux 2004) or their mediatic distribution can reveal public support for or resistance against them (Lisle 2011). The idea of how images resonate in IR is praiseworthy indeed.

---

[2] Here, I recommend the reader to listen to the song "Turn, Turn, Turn" by Byrds: https://www.youtube.com/watch?v=snZKnES4ng4

However, it is unclear in which directions and in general how social reactions to images contribute to the Discipline. The visual turn does not propose any precise method or theory (Williams 2018). Consequently, who gets what message from these photographs is unknown.

2.2   Game Theory in Visual Turn

Visual turn omits game theory. Yet game theory can expand the scope of the visual turn as a powerful methodological tool to analyze interactions. The omission is mostly due to the belief that there is no place for emotion, feelings, and aesthetics in game theory. This belief is wrong, however. Game theory does not exclude the role of subjective feelings such as fear of being cheated or joy of altruism (Osborne 2004, 27–28; O'Neill 2001; Fearon and Wendt 1992). The exclusion of game theory prevents the visual turn to study and dwell into interlocked subjectivities, interpretations, and meanings using a rigorous and formal language.

In general, VSGs come in two types depending on whether players have common or conflicting interests with respect to the type of the sign they agree upon. If a sign comes at the expense of the other on the basis of diametrically opposed preferences, then the VSG is zero-sum: it models a strict conflict of interest between players. A zero-sum VSG generates no chance of coordination, because players together cannot reach an outcome with better values for both of them. "A player cannot win unless he or she wins from the opponent; a player cannot lose unless there is a corresponding gain for the opponent" (Davis 1983, 44). Players having diametrically opposed interests and preferences over interpretations of the picture reach payoffs that sum up to zero. If players' preferences are not diametrically opposed, then they are involved in a nonzero-sum VSG where they have identical or mixed motives producing incentives to accept a common interpretation and might find a way out of the strict conflict by reaching a compromise or a mutually beneficial outcome.

Therefore, any equilibrium of a zero-sum VSG indicates that what one interpreter loses is what the other wins. A VSG can be nonzero-sum when players do not have diametrically opposed preferences. Players of nonzero-sum (mixed-motive) games have motives in which cooperative and conflictive incentives co-exist. Such motives generate interaction outcomes which are not necessarily at the expense of each other. Hence, mixed-motive VSGs involve mutual advantages for players in equal or unequal terms.

A short discussion and a simple example are instructive to illustrate the richness to discuss meanings of Abu Ghraib pictures in game-theoretic terms. The example is about two interpreters who interact to interpret the same Abu Ghraib photograph. The interpreters are assumed to be strategically interdependent: each must think about what interpretation the other concocts for the photograph because they cannot reach a meaning of the photograph unilaterally. They are not the sole deciders of what meaning the photograph represents.

What would the strategies of the players be? Suppose that each player possesses two strategies that correspond to an interpretation of the photograph as meaning of a Western, Christian victory in the Middle East against Islam, Muslims, or an interpretation of the photograph as the sign of the apogee of Western downfall in ethical terms. Players have no common ground of agreement about the meaning of the photograph; their preferences are diametrically opposed. Thus, no room for coordination can ever exist between them.

These suppositions produce two interacting players possessing two strategies each. Thus, we now have a 2×2 game. Furthermore, players are assumed to select their strategies simultaneously. Players opt for an interpretation without knowing what interpretation the other player has selected. The game is at strategic level specifying players, strategies, and preferences over the outcomes, that is, players' payoff functions. An immediate question then arises about how the two interpreters interact if they have diametrically opposed preferences over meaning they attach to Abu Ghraib pictures, or, equivalently, how the two interpreters interact by doing her best to hurt or overcome the other? Game theory answers this question through a zero-sum game model; interpreters select strategies to the detriment of the other. One interpreter, that is, a player, might prefer to interpret the photograph as a Christian victory over Muslims while the other as a Christian defeat facing Islam. Let's call these interpreters as Row and Column, respectively. Row and Column are supposed to possess two strategies corresponding to their interpretations, that is, meanings they construct about the picture. For Row, R_1 is equivalent to the strategy of "Abu Ghraib picture means Christian victory over Muslims" and R_2 is equivalent to the strategy "Abu Ghraib picture means the apogee of Western downfall in ethical and religious terms." Similarly, for Column, C_1 is equivalent to the strategy "Abu Ghraib picture means Christian victory over Muslims" and C_2 is equivalent to the strategy "Abu Ghraib picture means the apogee of Western downfall in ethical and religious terms." Players obtain inverse satisfactions; the parameter a in $R_1 C_1$ and $R_2 C_2$ cells measures players' diametrically opposed preferences on those outcomes while any disagreement generates no utility for both players. In each cell, the first utility is Row's and the second utility is Column's.

What will be the solution of the game?

The game is zero-sum because there is a strict competition between Row and Column who have diametrically opposed preferences, and, therefore, $a + (-a) = 0$ and $0 + 0 = 0$; the sum of utilities in each cell of the game matrix is equal to zero.

A quick glance at the matrix tells the story: the upper-left cell displaying players' utilities if Row selects his strategy "Abu Ghraib picture means Christian victory over Muslims" and Column selects her strategy "Abu Ghraib picture means the apogee of Western downfall in ethical and religious terms." The upper left cell gives the best outcome for Row and the worst for Column. Similarly, the lower-right cell is the best outcome for Column and therefore the worst outcome for Row. This is the outcome of both players interpreting the Abu Ghraib picture as the apogee of Western downfall in ethical and religious terms. There is only one strategy profile that is a Nash equilibrium, namely, (R_1, C_2), corresponding to "Abu Ghraib picture means Christian victory over Muslims" and "Abu Ghraib picture means the apogee of Western downfall in ethical and religious terms." Both players insist on their interpretation; they have no incentive to deviate from their preferred interpretation to the alternative interpretation because doing so implies that they accept the interpretation of the other.

In fact, against C_1, R_1 generates a utility of $a > 0$ but R_2 generates only a utility of 0. Thus, R_1 is the best reply against C_1. Similarly, against C_2, R_1 generates a utility of 0 but R_2 generates only a utility of $-a$, the worst outcome for Row. Thus, R_1 is the best reply against C_2 for Row. Hence, regardless of Column's choice, R_1 is strictly better than R_2 for Row. This means that R_1 strictly dominates R_2. Consequently, Row chooses R_1 to satisfy his preferences. Column is in a symmetric position as Row. Her strategy "Abu Ghraib picture means the apogee of Western downfall in ethical and religious terms," that is, C_2, generates strictly higher payoffs compared to her strategy C_1 regardless of Row's choices. This means that C_2 strictly dominates C_1. Row and Column know the matrix, and each knows that the other knows the matrix and so on, therefore the game is of common knowledge, each knows that the other has a dominant strategy. The equilibrium is the strategy pair of (R_1, C_2) assigning both players zero payoffs. There is no chance that Row and Column coordinate choices agree on the same interpretation of the picture.

2.3 Structural Constraints

Waltz (1979, 8) defines that "a theory is a picture, mentally formed, of a bounded realm or domain of activity… a depiction of the organization of a domain and of the connections among its parts." Structure is Waltz's main theoretical tool: "to define a structure requires ignoring how units relate with one another (how they interact) and concentrating on how they stand in relation to one another (how they are arranged and positioned)" (Waltz 1979, 80).

Waltz's words are reminiscent of French painter Georges Braque's view of instead focusing on objects individually, one has to focus on how these objects stand with respect to each other. Braque's view and Waltz's structure conceptualization are congruent. According to Braque, if they are taken individually, objects are not meaningful unlike their placements among other objects which generate aesthetic meanings. In Braque's words: "objects exist only as far as their mutual harmonious relationships, as well as those between the objects and myself, are concerned. Whenever this harmony is achieved, a sort of intellectual nonexistence is attained that makes everything possible and just" (Braque 1997, 6). Waltz argues that how states are placed toward each other in terms of resources they control, that is, structure, constrain and create a disposing force on the interacting states. Structures "explain why different units behave similarly, and despite their variations, produce outcomes that fall within expected ranges" (Waltz 1979, 72–74).

Structure has two elements according to Waltz: anarchy, that is, the inexistence of an overarching authority above sovereign states, and the distribution of capabilities across states that is the way states are positioned with respect to each other. Anarchy is the principle of states' arrangement and a constant because there is no authority nullifying states' freedom. A state's signing an international agreement does never prevent the state's actions from violating the clauses of that agreement if the state values the agreement as harmful against its interests.

The distribution of capabilities across states indicates states' relative positions and therefore their powers toward each other. "Structures denote states' arrangements. To define a structure requires ignoring how units relate to one another (how they interact) but concentrating on how they stand in relation to one another (how they are arranged or positioned). How units stand in relation to one another, the way they are arranged or positioned, is not a property of units. The arrangement of units is a property of the system ... anarchy is the principle of units' arrangement" Waltz (1979, 81). The system view of IR reflects Braque's view of the arrangement of objects as a central feature of paintings.

Waltz's principal claim is "to show how the two levels operate and interact, and that requires marking them off from each other. One can ask how state A and state B affect each other, and proceed to seek an answer, only if state A and state B can be kept distinct. Any approach or theory, if it is rightly termed 'systemic,' must show how the system level, or structure, is distinct from the level of interacting units." Waltz' view being parallel to Braque's implies that objects gain meaning by their placements relative to each other.

The Fig. 2.1 Waltz draws illustrates the placement and the reciprocal relationship between structure and interacting units, that is, international relations (Waltz 1979, 40). The figure consists of two rectangles and two arrows: the top rectangle represents international structure and the bottom one represents interacting units. The two are connected by two arrows. One arrow goes from the top rectangle to the bottom one. The other arrow goes from the bottom rectangle to the top one. The figure illustrates how Waltz's mind gets in contact with IR. It reveals and fixes the mental meaning of Waltz's main theoretical proposition as IR affecting international structure and IR being shaped back by international structure. Structural Realism demonstrates how international structures and interacting units interact corresponding to structural constraints that cannot be explored by isolated structure and state interactions. The arrows bind both elements.

What will be the solution of the game?

		Column	
		C_1	C_2
Row	R_1	$a, -a$	$0, 0$
	R_2	$0, 0$	$-a, a$

Fig. 2.1 A zero-sum VSG

The two arrows in the figure depict structural constraints shaping and shoving states' interactions. The problem is to prove the existence of the constraints and their intensity. The key question to be answered is: do structural constraints exist? Waltz does not specify the direction and the magnitude of structural constraints. He only claims that they "encourage states to do some things and to refrain from doing other" (Waltz 1997, 915). Structural constraints are assumed to exist; they are unobservable. One can then ask whether the concept of structural constraints corresponds to phlogiston, a substance burning materials lose once used to explain combustion? (Derry 1999, 185). We now know that phlogiston does not exist, oxygen is the necessary condition for combustion. Waltz assumes that structural constraints exist to advance explanations at system level by taking them as primary forces. Will structural constraints share the same fate as phlogiston?

We have to ponder about the meaning of observations to answer the question. The central problem is to define what an observation is. In biology, for example, one can observe viruses by an electron microscope or computerized techniques. In international politics, for example, one can observe a war as an outsider or as an insider living under horrendous conditions of human suffering. Are these observations distinct from each other? They are distinct of course. The question then becomes: what type of international political observations establish the existence of structural constraints? The debate between scientific realism and constructive empiricism allows clues for answers to these questions (Chakravartty 2007; Churchland 1979; Churchland and Hooker 1985; Van Fraassen 1980).

Scientific realism as philosophy of science doctrine "largely unacknowledged by political scientists" (Wendt 1987, 336) does not distinguish between observable or nonobservable forces. Do structural constraints refer to real but unobservable forces independent of mental workings of picture making? Do they vary across international systems regardless of how Waltz postulates them? These questions become central in scientific realist terms. According to scientific realism, structural constraints' existence does not depend upon our thoughts or language we use to describe them. If structural constraints mental pictures imply produce successful explanations of international politics, structural constraints can be argued to exist; not otherwise. Thus, scientific realism supports Waltz's conjecture of structural constraints provided that they are demonstrated to shape and shove international interactions. No such demonstration is given yet.

2.4 Scientific Realism, Phlogiston

Scientific realism claims that a precise distinction between observable and unobservable entities is impossible to draw (Okasha 2002, 66). Wendt (1987, 351) states that "Neorealists might be seen as scientific realists to the extent that they believe that state interests or utilities are real but unobservable mechanisms which generate state behavior." Structural constraints belong to the set of unobservable mechanisms as Wendt indicates. The problem is to transform Wendt's assertion into "Neorealists might be seen as scientific realists to the extent that they *explain* that state interests or utilities are real but unobservable mechanisms which generate state behavior." Structural constraints can be argued to exist according to scientific realism as long as they explain international political events.

In the context of phlogiston versus structural constraints, one must separate ontological assumptions from explanations. For example, if a scientific realist travels back in time when phlogiston is not observable but assumed to explain combustion, then she would accept that phlogiston exists. Therefore, scientific realism can support structural constraints' real existence only if they explain how structure shapes and shoves states' interactions. Hence, to reject structural constraints' existence, there must be a proof that structural constraints do not explain constraints over states' interactions. Short of such a proof, scientific realism can only imply that structural constraints exist independently of our beliefs, and perceptions. Waltz creates the concept of structural constraints; he does not discover them. Thus, his approach can be qualified as compatible with scientific realism as long as he explains variations in states' interactions across different structures.

Nevertheless, structural constraints can become observed forces in the future. It is possible to detect, observe, or discover unobserved entities later (Harré 1986, 57). But is it necessary that structural constraints are observed? The answer is no. No one saw molecules but explanations are built upon their existence. It is difficult to draw a precise distinction between observables and nonobservables. Hence, scientific realism does not eliminate explanations based on unobserved structural constraints. This is exactly where constructive empiricism deviates from scientific realism (Van Fraassen 1980). If, for example, central states are observed to spend efforts to prevent alliance between the wing states, then the scientific realist view uses these observations as grounds and supports to explanations and proofs of structural constraints' existence. Constructive empiricist view would take these observations as supporting the adequacy of explanations only unless one has the proof of the existence of these constraints. In short, the observations do not constitute the final proof of structural constraints for constructive empiricism. To wit, constructive empiricism does not believe in God unless God is seen.

Some examples of triadic international structures materialize Waltz's claim. Caplow (1959) offers eight possible power distributions in such systems:

1. $A = B = C$.
2. $A = B > C$.
3. $A = B + C$ and $B = C$.
4. $A > B = C$ and $A < B + C$.
5. $A > B = C$ and $A > B + C$,
6. $A > B > C$ and $A = B + C$.
7. $A > B > C$ and $A < B + C$.
8. $A > B > C$ and $A > B + C$.

Each power distribution represents three states' relative positions toward each other corresponding to international structure box in Waltz's figure. The concept of structure stems from the idea that "units differently juxtaposed and combined behave differently and interaction produce different outcomes" (Waltz 1979, 80). Therefore, each picture corresponds to a structure generating constraints that reduce the range of state interactions (Waltz 1979, 58, 73, 81). To expand Waltz's proposition, imagine, for example, that in each triad all states are arranged as sharing common border with each other or not so that one state is placed geographically between two others, or one state is away from two states that are contiguous to each other. Structural constraints vary in magnitude across these systems. Thus, Braque's view about the meaningfulness of objects stemming from objects' relative placements to each other now transforms into states' placements relative to each other across these eight distributions and expands Waltz's structure concept. States' geographic positions enrich Caplow's list of eight distribution of capabilities structures. The addition of geography adds a new structural dimension and demonstrates a parallel between Braque's view of abstract art and abstract thinking in IR in structural terms.

The first power distribution of A = B = C yields four pictures of international structures where all states are distant from each other, all share a common border, one is distant but the two are contiguous, and one is taken between the other two. If the power distribution is A = B = C, structural constraints that shape and shove three states' interactions seem to be of equal magnitude for states A, B, and C if they are all contiguous or discontiguous. However, if one of them is distant to two others that share a common border or one of them is placed between the two others, structural constraints vary while the power distribution does not change. Structural constraints change for each state having equal power position relative to other two others. Comparing, for example, A = B = C and A > B > C and A < B + C, one can assume that A, B, and C are subject to different constraints. In the former structure, each state is subject to identical constraints under the condition of equivalent geographic positions. If the structure is A > B > C and A < (B + C), then B and C incline to form a common front against A. States B and C are not as free in alignment choices as in A = B = C structure.

Questions about structural constraints multiply as result. What about if the strongest or the weakest state is distant while the other two share common borders? What if the strongest state is taken between the other others? Any answer or answers to these questions do not produce explanations of inquiry-independent phenomena on the basis of empirically proven structural constraints but constitute conjectures about structural constraints' existence and explanations they allow to make. It is possible to think that the structure of A = B > C refers to two equally powerful states and a weak one. If A and B are contiguous and C is distant and if A or B is contiguous to C, but they are apart from each other, structural constraints change corresponding to a structural explanation (Harré 1988, 132). Consequently, direction and magnitude structural constraints vary across all possible structures yet, without being observed, they have to generate successful explanations.

2.5   Criticisms of Structural Realism

IR scholars still discuss the meaning of theory including Waltz's definition of it. A leading journal in the field, consecrated one of its issues to explore answers to the following question: "The End of International Relations Theory?"[3] By certainty, such questions will also be asked in the future given the nonstop turning of the Discipline. If one offers a theory, attacks are immediate in the warzone of the Discipline albeit principal questions range from what would one expect from a theory, what is change, at what level can an IR theory explain change? to what is state? what is war? what is peace? The last three questions are reminiscent of Gaugin's canvas titled "Where do we come from? What are we? Where are we going?" painted in 1897–98.[4] The painting can be argued to picture the current state of the Discipline.

IR theorists reach a consensus in their criticism of Waltz's theory to a saturation point. More criticism would only mean beating "a dead horse" and "they are done to death" (Donnelly 2019, 905). The criticism misses many marks, however. First, IR scholars resist simplification and abstraction. They tend to argue for theoretical complexities to oppose simplicity of theories and therefore theoretical fundamentals once a simple theory is offered. If a simple theory is offered, criticisms go in the direction of the inclusion of empirical entities and connections without investigating what theory implies in its simple form and level of analysis. Second, the theoretical irony is to insist that Waltz should not exclude individual characteristics and traits of states (Donnelly 2019, 904). Yet the suggestion of the insertion of state-level traits is in direct contradiction with Waltz's aim of offering an IR theory at system level. State-level traits have no place in a system-level theory. An insertion of states' attributes runs against Waltz's aim of reaching an abstract theory through simplification as well. Similarly, if, for example, one offers an IR theory making states' attributes as central, the question would become why the theory does not take system-level attributes into account. There exists a strong resistance to explore implications of a simple theory in the Discipline indeed.

There is no system-level theory that generates explanations of events running against Waltz's theory. There is no other book than Waltz's *Theory of International Politics* that constituted a discussion source in the Discipline qualified as "nothing but a set of references to Waltz's book" (Waever 2009). Innumerable references to Waltz's book prove that the Discipline's fundamental problem is the lack of understanding of what a theory means and the inexistence of a common scholarly language. Unsurprisingly, much earlier criticism has targeted Structural Realism as being a static theory. Ashley (1986, 265) holds that Waltz has offered a theory which does not explain change. Explanations of change however require the use of mathematics constituting a powerful tool to probe dynamics of any sorts covering international politics. Verbal analyses lack the rigorous quality of mathematics.

---

[3] *European Journal of International Relations*, 19(3), 2013.

[4] The painting, oil on canvas, is in the Museum of Fine Arts, Boston.

Compared to mathematical studies of change, verbal discussions of change prove to be inadequate and fall short of breath in imposing meanings to conditions of change, that is, interpretations. Thus, there are dynamic models of IR like the arms race model of Lewis F. Richardson but no dynamic IR theories that explain change mathematically like those one encounters in physical sciences or in economics. A dynamic theory should deal with change over time and such changes cannot be studied in plain words but by differential equations; otherwise, how a theory explain change remains a mystery. For example, Richardson's arms race model is run by ordinary differential equations. The model and its extensions expressed in differential equations are studied in detail by Dina A. Zinnes (1976). Yet these models cover one aspect of IR only; they do not constitute an IR theory at system level as the one Waltz offers.

Waltz offers two concepts of change in IR: if the organization principle of international politics, that is, anarchy, disappears, then there is change of the system; otherwise, if distribution of capabilities across states change, then there is a change in the system (Waltz 2001, 1993). Changes in or of the international system can occur for any reason which are not of any theoretical interest as they lie outside the scope of the theory. How could one expect when or the exact date, time, decade of the international system being not composed of states but by some other units, say multinationals and firms possessed by some powerful men? How would one expect Waltz's theory to explain such a global transformation? An answer would come from imagination and fantastic movie scenarios about the change of the system. Think about futuristic movies such as Blade Runner (both the old and the new versions), Outland, Alien (with all its versions) and others centering on colossal mining firms sending ships in space looking for the extraction of precious metals, minerals in the solar system and beyond and perhaps extraterrestrial life forms. These science-fiction movies tell stories about changes of international systems where states are finally gone, and, therefore, it no longer make sense to talk about states and therefore about their co-existence corresponding to the condition of anarchy.

There are other types of criticisms of Structural Realism such as Structural Realism ignoring social elements (Wendt 1999, 1992) and it is ahistorical (Schroeder 1994). Postmodernists and poststructuralists criticize it in discursive terms as it conceals fundamental power-knowledge relations in the Discipline. Their main argument is that any realist theory of IR offers pictures of warmongers and generates self-fulfilling prophecies of inter-state conflicts (George 1995). The fact that the Discipline constitutes a warzone is already mentioned. The last charge sounds like an artillery attack or an exploding bomb in the battlefield of competing IR theories and methods. So, there is no change of the ongoing battle or prospects of peace in the Discipline. None of the criticisms listed concentrate on unexplored structural constraints. If real-world observations cannot establish the existence of structural constraints, then a painting can be of help to sense them.

On the other hand, attacks on Waltz galore. Still another attack is that Structural Realism is based on a positivist philosophy of science. Positivism rejects mental pictures, and unmeasurable forces. A positivist or a neo-positivist doctrine of philosophy of science would aim at establishing mental pictures as outputs of data measurements and correspondence rules converting theoretical entities into empirical observations. Waltz does not spend efforts in these directions. He believes that the human mind is at the center stage of structural realist thinking and therefore defines a theory as a "picture, mentally formed, of a bounded realm or domain of activity" (Waltz 1979, 8). Waltz searches for mechanisms of nonobservable nature but those which are always present and materialize occasionally. He insists on, for example, structure as an analytical category, not as a real one. Waltz's anti-positivist stand is given in his own words: "I emphasized that much in the present seems to contradict predictions I make. But then, I did not write as a positivist or an empiricist" (Pond and Waltz 1994, 194). If one reads Waltz's book *Theory of International Politics* and believes in Waltz's sincerity, or own words, or both, then one should take Waltz as being no positivist. Unsurprisingly, Waltz's words do not save him from being labeled as a positivist (Ashley 1986; Cox 1986; Keohane 1986). Waltz rejects the view that all knowledge comes from experience and observations. Unobserved structural constraints support Waltz's view of structures as constellations of states that shape and shove international interactions hinting at an anti-positivist philosophy of science Structural Realism is founded upon. Yet for many IR scholars Waltz assumes rationality of states. This is understandable as the Discipline is notorious in creating artificial discussions and nonfinalization of discussions so that another discussion starts before one is complete avoiding healthy communication.

Three decades after the publication of Waltz's book in 1979 an IR theorist finally remarks that Waltz makes a strong distinction between what is real and what is not: "In modelling a theory, one looks for suggestive ways of depicting theory, and not the reality it deals with. The model then presents theory, with its theoretical notions necessarily omitted, whether through organismic, mechanical, mathematical, or other expressions" (Wæver 2009, 204).

The sociological IR theory proposed by Wendt rooted in criticisms of Structural Realism posits a multiplicity of anarchies instead of a unique constant one. Both concepts are worth to be examined because Structural Realism and Constructivism constitute two central IR theories. They need to be examined by the help of VSGs given the controversial and poor state of communication in the Discipline.

2.6 Dynamic Nonunique Anarchies

Waltz defines anarchy as the organization principle of international politics and the coexistence of sovereign units. Contrary to Waltz, Wendt offers a sociological perspective on the basis of a rejection of Waltz's assumption of functionally undifferentiated states having the unique task of perpetuation of their sovereign existence. Waltz argues that material distribution of capabilities bends states' interactions together with anarchy, Wendt in turn argues that cultures bend material distribution of capabilities across states. Wendt argues that structures of international systems are social, that is, states are functionally differentiated under changing identities and interests through their interactions qualified as "practices." Social structures are ideas fused with material capabilities so that it is impossible to isolate material capabilities and ideas.

Wendt keeps material capabilities of Structural Realism and adds some sociological traits as new structural variables states' practices produce. States' practices constitute the origin of what each state expects from others so that states share knowledge and understand each other through practices. Material capabilities cannot constitute the sole determinant of structures of international systems in the presence of states' shared knowledge and expectations from each other. Changes in these expectations produce changes in states' intersubjectivity and different anarchies as social structures. Therefore, Wendt (1992, 1999) argues that "anarchy is what states make of it;" anarchy is a social construct (Wendt 1992, 395), that is, anarchy is a culture (Wendt 2000). There cannot be a unique anarchy in an international system. The dynamic nature of nonunique anarchies follows from ongoing international interactions generating cultures as "social structures exist only in process" (Wendt 1995, 74). The result is a multiplicity of dynamic anarchies.

Structural constraints are of social character in nonunique dynamic anarchies in Constructivism. Social structures shape and shove states' interactions unlike strictly material structures. For example, if states share a common belief that they are "friends," that is, they subjectively know each other as friends, then security of self loses its priority: cognitive variations bend the effect of material capabilities on states' interests. States can then take others' securities as if they are their own security. Thus, according to Wendt, states' uncertainty about intentions of others and states' expectations that war can occur anytime and therefore they must rely on own capabilities, that is, self-help, is not states' destiny contra Waltz. Anarchy does not necessarily lead to a self-regarding behavior implying a culture of state-eat-state competition. States can develop friendship in their interactions as their practices with each other generate intersubjective knowledge positing each state as a friend. There is no constant Hobbesian state of nature dominating international politics. Yet, there is no rule preventing friends to become enemies due to their interactions with each other in the future. A chance event like an unexpected misunderstanding, a misperception, or any other ideational factor can make anarchy among friends to shift into the opposite direction with an anarchy among enemies form ultimately.

2.6.1 Categorization of Anarchies

Wendt (1992, 400) proposes three types of anarchies (cultures) among states: competitive (Hobbesian), individualist (Lockean), and cooperative (Kantian). The Hobbesian anarchy is a "competitive" one where states' intersubjectivity is evolved to a state of nature so that each state views other states as enemies. It corresponds to Waltz's conceptualization of anarchy where states' uncertainty about mutual intentions generates a social setting where war can occur at any moment. States identify each other as friends caring about mutual and ultimately about collective global interests in the Kantian anarchy. States' practices indicate an altruism which can be expressed as states' mutual perceptions and views of "your gain is my gain; your cost is my cost." The Lockean anarchy is "individualist." It is a mixture of the first two. States are mostly concerned with themselves and care about individual gains. They do not compare their gains with those gains other states reap through practices like in Hobbesian anarchy. States become open to cooperative endeavors being not fixated on material resources. Hence, each anarchy varies in its social trait and intensity, for example, the individualist and the competitive anarchies can converge but never become identical. There could exist cooperative systems ranging through states' identification with each other "from the limited form found in 'concerts' to the full-blown form seen in 'collective security' arrangements" (Wendt 1992, 401). Hence, states' practices producing distrust can lower the level of friendship and amity in a cooperative anarchy making it to approach an individualist anarchy. Similarly, changing identities and therefore interests of states indicating that, through practices, they are mostly concerned about their own security can make a competitive system to approach an individualist anarchy. Individualist anarchies can similarly approach toward cooperative or competitive anarchies or both. All social transformations of anarchies indicate possible directional changes in states' identities and interests over time implying a "continuum of anarchies" (Hopf 1998, 174). Thus, anarchies are multiple, dynamic, as they shift from a Hobbesian one to a Lockean or a Kantian one or vice versa due to changes or unexpected developments in states' intersubjectivity. There exists a range of numerous anarchies not only three.

2.6.2 Changing Anarchies

A point most IR theorists miss is that Waltz discusses socialization as a process through which states are mutually affected by their friendship and hostility. Waltz (1979, 74) remarks: "Consider the process of socialization in the simplest case of a pair of persons, or for that matter of firms or states. *A* influences *B*. *B*, made different by *A*'s influence, influences *A*." Thus, structural logic Waltz mentions does not imply a multiplicity or a classification of anarchies; instead, the logic implies how the type of a relationship can evolve through mutual transactions and network. If, for example, State *A* becomes friendly with State *B*, then State *B* might reciprocate state *A*'s friendship to some extent; it is impossible that no impact whatsoever would occur on State *B*'s actions towards State *A*. The same would apply if State *B* becomes hostile for some reason toward State *A*. State *A* would note that hostility and would adapt its interaction with State *B* accordingly. The difference between socialization conceptualization of Waltz and nonunique dynamic anarchies of Wendt surfaces at this point: Waltz does not note ideational factors in socialization unlike Wendt. Waltz simply notes the mechanism of structural functional logic, that is, how relations evolve through interactions without discussing the effect of states' intersubjectivities on IR through diplomacy and discursive practices.

Wendt's perspective makes structure a variable. Structural constraints assume innumerable directions unlike in Waltz's graph where one arrow emanates from structure in direction of interacting units and the other emanates from interacting units to structure. The possibility of innumerable changes of structural constraints makes one wonder about whether there are social conditions where practices permit different cultures to intersect in Wendtian perspective of anarchy. For example, two distinct international systems one characterized by Hobbesian and the other characterized by a Kantian anarchy can start to interact. The union of the two systems then may trigger a Lockean one through mixtures of states' intersubjectivities. It is possible that living in a Kantian culture some proportion of states may perceive as if they exist in an individualist. The result, depending on the proportion of those states, would be that states start to shy away from the altruistic views of each other to some degree and be more concerned of their absolute gains. Such cultural transformations can be interpreted as evolving mixtures of anarchies over time.

2.6.3 Evolutionary Games and the Continuum of Anarchies

Evolutionary games study such transitions by its equilibrium concept of evolutionarily stable strategies (ESS) (Maynard-Smith 1982). Hence, a meeting or diversion of anarchies due to changes in conditions of states' shared knowledge, intersubjectivity, and therefore identities can be studied by the method of evolutionary games enriching Constructivism. Wendt (2000, 174) makes an evolutionary game statement by affirming that "*If* a critical mass of states position each other as rivals, the system will acquire a Lockean structure with certain hypothesized dynamics; *if* a critical mass of states begin to treat each other as friends then this culture will be transformed into a Kantian one; and so on."[5] Wendt's hypothesis finds a home in the concept of ESS according to which an entry of states having a different worldview compared to the rest of the population of states can be imitated depending on their number and therefore the previously dominant culture can be replaced by a different one. Any point on an evolutionary trajectory leading to a particular ESS corresponds to how proportions of states adhering to alternative cultures, like Kantian versus Hobbesian, Hobbesian-Lockean, and Kantian-Lockean change and demonstrate how the evolution of cultures evolve in an international system so that states alternatively adhere to roles of enemy, rival, and friend.

Evolutionary game theory does not need the assumption of rationality to derive consequences of its axioms. Thus, the correspondence implies that no state must be assumed as an actor who possesses clear objectives and means to achieve these objectives at a minimum amount of cost, that is, no state needs to be conceptualized as a rational agent in a context where cultures evolve. States as agents and as interacting units with changing fitness across anarchies demonstrate that their number

---

[5] Italics in the original.

adhering to a type of anarchy and mutants, that is, the number of states who may enter the population but who adhere to a different kind of anarchy summarizing the structure of the system, constitutes a support to Wendt's statement of "It is impossible for structures to have effect apart from the attributes and interactions of agents" (Wendt 1999, 12). The relationship between the size of states happy in one anarchy and those who are not constitutes a material dimension: the number of states. States' fitness becomes a function of resources they collect and also legitimacy of whatever norms ruling the system. Nevertheless, states being satisfied or not are ideational attributes of agents independent of material dimensions. Hence, the evolution of anarchies depends on both material and ideational components of international systems. Comparing with Waltz's structure conceptualization, Wendt's conceptualization of structures of systems allows for sources of change not only depending on the distribution of capabilities but also depending on the distribution of ideas showing that structural changes do not have a single but a dual source in contrast with Structural Realism.

Wendt (1999, 20) asserts a relationship between material power and interests and anarchies, that is, social cultures: "the effect of material power and interests depend on the social culture of the system." The implication of Wendt's claim is that as the culture of an international system varies across Hobbesian, Lockean, and Kantian anarchies, the impact of the distribution of capabilities upon structural changes varies as well. Therefore, the shared knowledge among states in a system shapes the impact of material sources upon changes of social culture. The question then becomes what changes of shared knowledge upon material powers and interests of states produce swift changes in anarchies. In Wendt's terms, can a transformation of "sauve qui peut" Hobbesian anarchy into a Kantian based on mutual cooperation and identity of interests among states be impossible or will it take a longer time compared to the transformation of a Lockean anarchy into a Kantian one? Evolutionary game theory can answer this question based on the size of the population where Lockean culture is adopted and the fitness of states in the Lockean culture and the size of those mutant states who adopt Kantian culture and their fitness among those who adopt Lockean culture. The imitation of the successful states in the Lockean culture according to Malthusian, linear, nonlinear, assortative matching dynamics can imply stability of the new culture or polymorphic ESS so that one group of states adheres to the old Lockean culture and the remaining group adheres to the Kantian culture.[6] There is no IR work studying anarchy transformations.

VSGs operationalized as evolutionary games can shed light upon coordination of IR theorists. A drip painting by Pollock can mean shifting and changing cultures as each color and direction of the paint meaning how changing distribution of knowledge across states transform the impact of the distribution of the distribution of capabilities on anarchies. Players' subjective evaluations and interpretations of paintings underlying players' coordination remain in the anti-realist realm of

---

[6] See Maynard-Smith (1982), Samuelson (1997), Friedman (1998, 1991) for alternative evolutionary dynamics. Güner (2017) uses evolutionary game theory to study balancing and bandwagoning in unipolar systems.

semiotics. Therefore, an ESS of a VSG refers to players' assessments on the evolution of a specific anarchy under the constraint of the specific painting. Players' interactions lead to stable meanings that do not exist independent of players' minds. Players' mental beliefs and ideas about changing cultures present an anti-realist view.

Similar to the condition of observability and existence of Waltz's concept of structural constraints, it is possible to ask what constraints anarchies Wendt conceptualizes generate. Do Kantian anarchies constrain states' choices so that they remain friends or do Hobbesian anarchies constrain states' choices so that they remain foes or do Lockean anarchies not constrain states' choices as they represent mixed intersubjective understandings among states? States' mutual expectations set states' identities as friends, foes, and mixtures of friendship and animosity across Kantian, Hobbesian, and Lockean anarchies respectively, but anarchies as socially constructed worlds do not remain fixed and evolve over time. A small perturbation in a Lockean anarchy can spawn a Hobbesian or a Kantian anarchy and therefore changes in structural constraints states face. Hence, the main difference between Waltzian and Wendtian anarchy understandings is that structural constraints do not change direction in Waltzian conceptualization unlike in Wendtian anarchies displaying higher levels of complexity through changes in states' practices with each other.

Changing anarchies on the basis of social norms and ideas constitute another ontological problem: do anarchy variations exist? And, in epistemological terms, how do we know that they exist? A claim that dynamic and changing anarchies do not exist reflects a view which reject empirical realities. If one supports the view that social interactions and international interactions cannot be separated from each other and that they are historical products, then observability chances reduce as multiplicity of interpretations increases.

2.7   Criticisms of Constructivism

Keohane's edited book "Neorealism and Its Critics" contains evaluations opposing Waltz's theory and also Waltz's responses to his critics. Likewise, Wendt responses to critical assessments of his book *Social Theory of International Politics* in a volume of the journal *Review of International Studies* published in 2000.[7] Keohane, one of the critics of Wendt, remarks that Wendt is after a demonstration of the view that the world is not entirely composed of matter and "the priority of the whole against parts," that is, anti-materialism and holism (Blackburn 2016, 294, 224). He adds that Wendt's claim is compatible with rationalism and can accommodate rationalism and anti-realism. Wendt's theory at system level, similar to Waltz's theory,

---

[7] *Review of International Studies (2000)*, volume 26, number 1. There exists more sources discussing Wendt's theory an criticizing it; it is impossible to review all these sources.

implies that states' foreign-policy actions are not in the scope of Constructivism. Therefore, Constructivism is not a theory of foreign policy. Yet Keohane criticizes Wendt's *Social Theory of International Politics* as the book contains "no propositions about state behavior." Thus, the criticism misses the mark.

Keohane further remarks that Wendt takes Waltz's *Theory of International Politics* and his model and target, but that Wendt's principal concern is "the role of ideas in world politics" (Keohane 2000, 126). Indeed, the distribution of capabilities leaves its paramount position in Structural Realism and the distribution of ideas replaces it in Constructivism. Constructivism constitutes a theory based on states' social consciousness in anthropomorphic terms. Keohane (2000, 126) in turn claims that Wendt's constructivist theory implies a duality of material forces versus subjective ideas and search for the primacy of ideas over material forces: "Creating this dichotomy is a little bit like arguing whether the heart or the brain is more fundamental to life." Wendt (2000, 168) counters this claim by arguing that his theory is after finding an answer to the question of "not what matters more but does the intersection of the two produce the outcomes we observe?" Hence, Wendt proposes inseparability of ideas and material forces in international politics, that is, there exists a mutual constitution of the two determinants of international politics.

Smith (2000) claims that Wendt mixes reason and cause. According to Smith, Wendt is mistaken in assuming that reasons can be causes; rationalists would greatly appreciate such a view (Smith 2000, 158). Qualifying constructivist perspective of international politics as "Wendt's world" he adds that: "In conclusion, let me comment on what his world look like. First it is a world very similar to the world of rationalist scholars. They will find much to admire about the book and much with which they agree." Wendt's view of reason-cause relationship due to Davidson (1963) who proposes that reasons rationalize actions. If a player in a VSG gives a reason why she believes that a Rothko painting constitutes a sign for Waltz's proposition, then it means that she rationalizes her choice. It is not sufficient just to note that the sign is evident, the sources of why she prefers that sign must be explained in detail. The rationalization must expose the attractive side of the sign, for example, a symbol, compared to icon and index. Once rationalization is exposed, explanation of the sign choice is explained.

The implication of Wendt's proposition is to be found in a drip painting of Jackson Pollock so that swirls and shifts of color can be interpreted as different points of such mutual constitution. A fundamental difference between Waltz and Wendt's theoretical approaches is that Wendt uses more complexities to arrive at an abstraction unlike Waltz. In the midst of theoretical differentiations and the tension between traditional and theoretical approaches there is a pressing need for a meeting of minds in the Discipline. The cursory overview of the Discipline demonstrates that coordination among IR theorists is critical for a common language to develop among them. The critical questions are how could IR scholars coordinate or fail to do so and how IR scholars achieve fruitful, efficient outcomes when they interact? VSGs generate answers to the critical question.

Chapter 3
Art Versus IR Theory

Abstraction connects abstract art and IR theories. The correspondence between the two realms stems from mental representation of essentials at levels of degree. An artist offers his/her subjective appreciation of worldly beauties in terms of colors, shapes, and the placement of these elements in a canvas and enjoys an upper hand in creating abstractions compared to IR theorists. IR theorists in turn analyze international political events using alternative approaches. However, an artist's freedom in the use of space, shape, color, line, and so forth constitutes definitely a larger freedom compared to an IR theorist who abstracts away some elements of international politics to reach bare essentials of an interaction.

3.1 Abstraction in Art and IR Theory

The more abstract becomes an artwork the fewer become essentials it contains that strike "a balance between the subjective and the objective" (Mondrian 1951). For example, an impressionist work by Claude Monet can be qualified as abstract but not as abstract as a painting by Paul Klee. A color-field painting by Mark Rothko can in turn be evaluated as more abstract than a painting by Paul Klee on the basis of colors, shapes, and placements of shapes in a canvas. Likewise, a comparison of how Waltz and Wendt define structures of international systems implies that Structural Realism is more abstract than Constructivism containing fewer elements to create a systems theory of IR. Here, we note Waltz uses a larger freedom of abstraction compared to Wendt. Constructivism becomes less abstract than Structural Realism which precedes it while Rothko and Pollock's artwork becomes more abstract than cubism that precedes structural abstraction. Thus, history of art and the history of the Discipline can evolve in opposite directions.

IR theories possess abstraction levels similar to abstract art but exclude forms abstract art possesses exposing subjective abstraction levels. They do not vary in terms of forms or colors but in terms of IR traits they select to explain and

© Springer Nature Switzerland AG 2023
S. Ş. Güner, *Art and IR Theory*, Mathematics in Mind,
https://doi.org/10.1007/978-3-031-32342-3_3

understand international phenomena. Structural Realism assumes that structures of international systems are determined by the co-existence of sovereign states meaning the principle of states' organization known as anarchy, the number of states, and the distribution of capabilities across states (Waltz 1979). The structure Waltz proposes is a "container concept" as defined by Arnheim (1969, 174) because it helps to identify an international system's basic attributes. Constructivism in turn defines structures of international systems in sociological terms such as shared knowledge, expectations, meaning, and practices (Wendt 1999). The sheer number of structural traits both definitions are based upon implies that Constructivism is less abstract containing more complexities compared to Structural Realism.

The difference between IR theories and abstract art in terms of abstraction levels does not prevent meaning making in theoretical domain of IR using art. IR theories are abstractions in linguistic form. Paintings are based on an artistic syntax which is not equivalent to verbal, linguistic syntax (Langer 1951, 86–89). Colors, figures, and shapes open far larger interpretive ranges than texts offer. Thus, IR abstraction is a subset of abstraction in art implying that the correspondence between the two realms is a one-way avenue: IR theories do not generate meanings for art, but abstract art does so for IR theories. Abstraction contours are larger in art compared to the discipline of IR theories. Artistic sophistication through senses using media is pitched at much higher level compared to forming IR theories. The revelation of senses and aesthetic appreciation producing essentiality of the thing abstracted in art is inexistent in IR verbal, linguistic theoretic thinking. Hence, IR theoretic thinking cannot fully emulate artistic sophistication in abstract terms. IR theories are incapable of reaching artistic forms spawning mental sensations of color, shapes, and the placement of shapes in an abstract painting. The good news is that the level of abstraction in IR theories being a subset of the level of abstraction in art permits and facilitates the migration of mental images from art to the Discipline.

The migration does not go in both directions. Rothko's words of "there is no such thing as a good painting about nothing…we favor the simple expression of complex thought" (New York Times, 1943, September 9th) are reminiscent of how IR theories constitute simplifications of complex international political events. Similarly, Rothko's color-pane canvases spawn mental sensations of Waltz' mental image. Therefore, the sheer quality of abstraction of IR theories and abstract art permits migration of mental images from the artistic realm to the linguistic realm of IR theories allowing humans to attach meanings to linguistic objects through interpretations visual semiotics allow and permit.

The use of semiotics in IR is supported by Paul Klee's dictum of "art does not reproduce the visible, it makes visible" (Klee 1920, 28).[1] The dictum implies for the Discipline that IR theories do not reflect the visible reality, empirical facts, and observations. Instead, they explain facts by revealing forces that escape careful, detailed descriptions. They serve to abduce unknowns that explain events as abduction functions through eliminating nonessential properties and keeping the essential

---

[1] "Kunst gibt nicht das Sichbare wieder, sondern macht sichbar."

ones interpreted as creating meanings (Peirce 1960a, b, c). Waltz's definition of theory as "a picture, mentally formed, of a bounded realm or domain of activity; a depiction of the organization of a domain and of the connections of its parts" seconds Paul Klee's view of art (Waltz 1979, 8). The definition implies that a theory does not reproduce visible interactions among states, but rather it makes international structural forces visible through nonphysical mental pictures.

Mental pictures require abstraction. IR theories are in essence abstract modes of thought that generate explanations of real-world events. They illuminate connections between mental entities and forces of international politics. IR theorists' attempts to descend to the bare essentials of international interactions parallel artists' attempts to capture the essence of an object representing something previously invisible. Artists' and IR theorists' efforts mutually correspond. An example is Piet Mondrian's painting trees by pruning wigs and branches step-by-step to arrive at simplified tree representations. The tree is no longer painted in detail but in a few stylized branches (Figs. 3.1 and 3.2).

Both paintings expose a rhythm in branches and the trunk, yet their abstraction levels vary. The Blossoming Apple Tree reaches a higher level of abstraction compared to the Grey Tree because, Grey Tree is closer to the objective, observed reality compared to the Blossoming Apple Tree. Mondrian achieves a higher level of abstractness and subjectivity in Blossoming Apple Tree by introducing pale colors that present an airy freshness, sense, and even the smell of blossoming flowers. Gazing at the Blossoming Apple Tree one is discouraged to get an orange peel smell, the artwork zeros in apples, their smells, and flowers. It is difficult to imagine that two people in front of the canvas think that each thinks about oranges rather than apples. The painting forms a locus of interpretive convergence. The form is more complex in the latter painting compared to the former one, but the level of abstraction is pitched at a higher level in the latter painting as pale colors communicate feelings and branches of the tree reach the ground and the underground. The higher number of branches serves the abstract idea of a tree covering the upper and the bottom parts of the canvas and becoming in union with the ground and the

Fig. 3.1 "Grey Tree" (1911) by Piet Mondrian. (https://medium.com/ signifier/piet-mondrians-tree-paintings-cef4ccac881. Accessed on 23 Oct 2021)

Fig. 3.2 "Apple Tree, Blossoming" (1912) by Piet Mondrian. (https:// medium.com/signifier/ piet-mondrians-tree-paintings-cef4ccac881. Accessed on 23 Oct 2021)

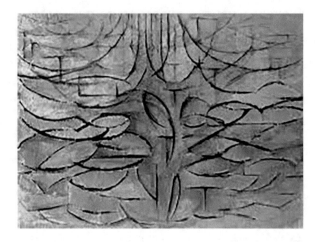

underground. There is no space between the ground, underground, and the top of the tree. The shape of the branches being similar in the imagined space of the ground and the underground implies that Mondrian does not differentiate between branches and the root of the tree. The union of the tree with the sky, ground, and the underground deciphers a holistic idea as the tree spans the whole space. This is where the higher abstraction level of the Apple Tree, Blossoming lies.

Therefore, a reduced number of branches is not sufficient to achieve abstractness in art. The idea of the tree unified with the sky, ground, and underground shows, in words of Mondrian, a higher balance between the subjective and the objective compared to the number of branches. The distinction of abstract art and IR theories comes exactly at this ideational freedom level provided by the intersection of objectivity of the outside world and the subjectivity of the artist. IR theories have a linguistic form; therefore, they do not possess the artistic freedom of the abstract representation of some complexity. An IR theory in its linguistic form cannot match Mondrian's use of colors and shapes. IR theories reach abstraction through simplicity of sparse international elements only. Abstraction in art can be achieved through subjective complexity an IR theory cannot reach. Yet while an IR theory cannot use colors, it can be interpreted as if it is colored in semiotic terms. A Rothko color-field painting, for example, can generate alternative meanings for Structural Realism through interpretation (Güner 2019) and some Pollock paintings do so for Constructivism (Güner 2021). Thus, an IR theory can be interpreted as if it is colorful.

Mondrian's abstractions echo IR theorists' efforts to simplify observations to manage their complexity and arriving at a parsimony level and explanations without taking note of every aspect of events. There are arguments in the Discipline running against theory development but supporting in-detail descriptions of selected cases. Morton Kaplan defines the cleavage in the Discipline as "Traditionalism versus Science" (Kaplan 1966). Mondrian's work evolving through step-by-step abstraction and Waltz's statement of "market structure is defined by counting firms; international political structure, by counting states. In the counting, distinctions are made only according to capabilities," constitutes parallel approaches in different realms (Waltz 1979, 99). Theoretical thinking requires abstraction in IR unlike the traditional approach. The tree and international political structure are both targets of abstraction: Mondrian gets rid of some branches and wigs; Waltz gets rid of individual and state-level traits retaining only those that are at the system level: the number of states and the distribution of capabilities across states. If Waltz looks at the international politics to assess the international structure, he counts the number of superpowers. This is of course an anathema for the traditionalist IR approach. If Mondrian looks at a tree to paint it in a simple form, he prunes some branches not all of them and not in their real, observed forms. One can imagine that an IR traditionalist advising Mondrian in his New York studio that additional squares and rectangles are necessary to create a "better" painting in colors other than the primaries of blue, red, and yellow. This example would summarize the mental approaches to theory and description in detail in IR. Hence, Mondrian's tree abstraction and Waltz's international political structure abstraction are comparable to some extent.

Total abstraction is possible in art but not in IR. To illustrate, the painting "The Black Square" by Malevitch is pitched at such an abstraction level no IR theory can imitate. It is impossible to pick a unique feature of international politics to concoct explanations. In his comic book titled "School is Hell," Matt Groening draws a figure of a college teacher who utters: "The nation that controls magnesium controls the universe!!!" and he qualifies the teacher as "the single-theory-to-explain-everything maniac" (Groening 1987). Malevitch's Black Square below would sound like that teacher's theory (Fig. 3.3).

Abstractness of art serves as a vehicle to attribute meanings and to interpret IR theories. Colors, shapes, and the placement of these elements used in a painting create mental possibilities of proposing metaphors or similes or other relations. They become cognitive statements making paintings symbolic vehicles for IR theories.

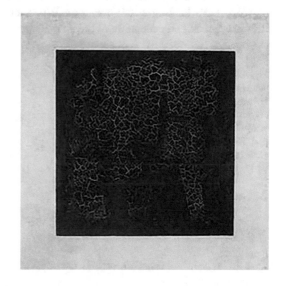

Fig. 3.3 Kazimir Malevich, Black Square, 1915, oil on linen, 79.5 × 79.5 cm, Tretyakov Gallery, Moscow. (https://en.wikipedia.org/wiki/Black_Square. Accessed on 24 Oct 2021)

Consequently, metaphors are to be found, easy to formulate yet difficult to appreciate interpretatively. The problem revolves around assessments of similarity, common elements of structural abstract paintings and theories as these assessments are pitched at aesthetic, sensual qualities of abstraction in art. The assessments can be infinitely many. Coordination, agreement of some feature of an IR theory, constitutes the heart of the communication problem.

3.2   Rothko-Waltz

There exists a subjective exactness between the proposition and the painting "Green and Tangerine on Red" by Rothko below divulges of a correspondence between them (Arnheim 1969, 116). The similarity lies in two colored panes in the painting and two rectangles in the graph connected with two arrows. The graph and the painting are visually connected representing alternative meanings of the connection. What matters in Waltz's theory is the structure like Rothko's work where the key is the juxtaposition of colors and colored areas occupy in the canvas. They constitute a parallel focus of abstraction. Differences in power become, transform into differences in colors and hues in Rothko's color field paintings. Had Waltz drawn circles instead of rectangles in his figure, a different painting would point to the exactness of the correspondence between the claim and the painting. The luminescence color interactions produce correct the view that SR is a static theory. It implies a dynamic view of structural constraints.

The painting assimilates the figure as abstract-abstract relations go from the artwork to theory by catching theoretical elements the figure misses to express (Fig. 3.4).

It is almost impossible to communicate and make meaningful statements about how one feels looking at the paintings. When an IR theorist faces a painting and tries to appreciate the meaning of the artwork, she feels sensations in her mind. The

Fig. 3.4 Green and tangerine on red. (https://www.phillipscollection.org/research/american_art/artwork/Rothko-Green_and_Tangerine.htm. Accessed 20 Dec 2021)

feelings are private, indescribable, and even ineffable. Subjective mental elements are strictly personal. However, when she communicates her interpretations of the paintings delimited by the IR theoretical statement and a variety of signs, she finds that the seemingly infinite area in subjective terms reduces to a considerable degree. This amounts to a two-step reduction: first, she is constrained to make a meaning, and second, she exchanges what she thinks about the sign connecting the artwork with another theorist who has her own subjectivities. The result is a further reduction of the domain of interpretations. The communication does not start with the two theorists exchanging ideas from scratch. Therefore, two IR theorists' mental interpretations based on reciprocal subjectivities produce limited and ordered signs.

The painting Green and Tangerine on Red offers soft and brushy borders of colored areas. Colors and the background interact with one another particularly around the edges. The painting resembles Waltz's graphs as the interacting rectangles in Waltz's graph generate sensations of color interactions. The resemblance relation between the painting and the graph corresponds to an analogy, therefore an iconicity relation. Consequently, the verbal and the graphical meanings of the graph get flesh and bones in the context of the painting. Rothko communicates powerful emotions through his art and Waltz communicates his principal proposition through his graph. The painting and the graph together yield a sound basis of interpretations on structural constraints. The latter gains meaning by the help of the former.

Rothko's works are not limited to subjective dimensions opened up by interacting colors and optical illusions they produce. They invite spectators for spiritual journeys. An IR theorist's spiritual journey in this context drives her to a deeper understanding of structural constraints Waltz posits to exist. The question could even become what interpretations and assessments of structural constraints could be reached through other Rothko color-field paintings. For example, one question becomes whether the dark, greenish area corresponds to structure and the tangerine part to interactions or vice versa. A selection of color-field paintings by Rothko on the basis of colors and their placements reveal iconicity and noniconicity relations.

There are Rothko paintings containing two colored rectangles. If the dark color on top and the light color on the bottom or if the dark color on the bottom and the light color on top, then meanings change. If the dark green area is interpreted as structure and the tangerine area is interpreted as interacting units, then there exists a full similarity between the graph and the painting. An iconicity of the painting for theoretical proposition relation exists. If one interprets structure as the fundamental force, and therefore dark area should be placed at the bottom not at the top of the painting, then iconicity relation disappears. An alternative Rothko painting would then become a better ground of interpretation. The full, physical similarity between the graph and the painting reappears but on the basis of a different painting. Therefore, an exploration of paintings by Rothko to make meanings of Waltz's proposition unlocks alternative paths enriching interpretations of structural constraints as appreciations of color interactions help to unveil verbal theoretical statements hide.

Could any color-field painting help to form signs for Waltz's proposition? The answer is affirmative. The painting "Untitled, 1972" by Ellsworth Kelly below is in

Fig. 3.5 Untitled 1972. (https://www.mutualart.com/Artwork/Untitled/267C2B7A9BCC8C03.
Accessed on 4 Jan 2022)

conformity with two colored areas superposed. One can compare her sensations of
color interactions by gazing at the painting and Rothko's Green and Tangerine on
Red. Naturally, senses will differ from one person to the other. Color interactions in
Ellsworth's painting are more difficult to grasp compared to Rothko's paintings
where edges are blurred, and one feels like colors mesh with each other. In a sense,
Rothko helps the spectator to be engulfed in color sensations unlike in Ellsworth
Kelly's Untitled, 1972. Consequently, Rothko's paintings are useful for making
meanings. They create existential experiences and universes of subjectivities and an
abstract language of feelings ranging from fear to loneliness. It is not an exaggera-
tion to qualify them as portals opening to universes unknown to the spectator
(Fig. 3.5).

3.3   Pollock-Wendt

Wendt offers a dynamic conceptualization of anarchy. Anarchy is a variable which
can change through practices. Pollock's drip-paint artwork helps to form signs of
Wendt's dynamic and nonunique anarchies. Interpretive associations of Wendt's
claims and selected Pollock paintings lead to mental discoveries enriching
Constructivism. Thus, one can take Pollock's any drip-paint work as compatible
with the rule of *aliquid stat pro aliquo*, that is, something stands for something else.
Players' interpretations of a drip painting by Pollock as it stands for Wendt's claim

are crested in players' minds. As result, abstract art of Pollock generates rich inter-pretive axes, anchors, and guides for Wendt's claim. Compared to Waltz's key vari-able of distribution of capabilities and therefore power across states Wendt brings attention to how states' interactions with each other amounting to social practices and therefore cultures states share with each other shape the level of conflict and cooperation in international systems. Thus, drip paintings represent changing cul-tures as a function of changing interactions among states through swift changes in direction and color of paint.

It is unlikely that all IR scholars agree that a correspondence exists between Pollock's art and Wendt's claims. Riegl's concept of Kunstwollen, that is, a dynamic drive targeting artistic orderings of the perceptual world, supports a parallelism between sign making through interpretations of Pollock's paintings as artistic order-ings of constructivist claims. This is not main question, however. The central point is to enrich and expand Wendt's claims. Pollock's drip-paint art constitutes one of innumerable art-IR theory correspondences and associations. There is an interpre-tive freedom in associations of art and IR theories. Hence, Pollock versus Wendt is only one of many other alternative associations.

Arnheim (1969, 35) notes that "... perceiving a work of art is not accomplished suddenly. More typically, the *observer starts somewhere*, tries to orient himself as to the main skeleton of the work, looks for accents, experiments with a tentative framework in order to see whether it fits the total content and so on. When the explo-ration is successful, the work is seen to repose comfortably in a congenial structure, which illuminate the work's *meaning* to the observer," and he adds that "visual perception, I tried to show, is not a passive recording of stimulus material but an active concern of the *mind*" (Arnheim 1969, 37).[2] It is possible to start observation from anywhere in Convergence as curls of dripped color cover the surface. The curls evoke or simulate a dynamic complexity of interest and identity formation and therefore anarchies. The multiplicity of color shifts and turns refers to the exclusion of Waltz's unicity of anarchy argument and to the exposition of the multiplicity of intersubjectivities describing anarchies. They locate the assertion of "self-help is due to process, not structure" (Wendt 1992, 394). States have changing identities and interests through those colored twists and turns. Convergence is not the unique pollock painting helping to interpret Wendt: Reflection of the Big Dipper (1947), Full Fathom Five (1947), Cathedral (1947), Number 1A, 1948 (1948), Lucifer (1947), One: Number 31, 1950 (1950), Lavender Mist: Number 1, 1950 (1950), Number 27, 1950 (1950), Number 13A,1948 (1948), Autumn Rhythm: Number 30, 1950 (1950), Number 29, 1950 (1950), Summertime: Number 9A, 1948 (1948), and Blue Poles: Number 11, 1952 (1952) are alternative drip-paint works of Pollock (Landau 1989). None being framed, each "appears more completely and modestly the artist's work" (Schapiro 1972–1973, 11). They can all help to interpret Wendt's conceptualization of social structures of international politics and visualize the opposition between Wendt's multiple dynamic anarchies and Waltz's unicity of

---

[2] Emphasis are mine.

Fig. 3.6 Convergence. (https://www.jackson-pollock.org/convergence.jsp)

anarchy. Pollock's drip-paint canvas of "Convergence" is an appropriate painting to demonstrate what could be learned from semiotics in interpretations of nonunique dynamic anarchies (Fig. 3.6).

The distinction between anarchies is impossible without a touchstone-like color or type of color shift or both. The painting permits to start anywhere on the canvas to start making meanings. One observes black, white, yellow, and orange movements some in the form of curls some not and also blue accents dispersed. If one identifies three anarchies Wendt describes with curly movements of different colors, then she can interpret them as taking different directions placing the painting in a semantic space of Constructivism. As a result, linguistic meanings of dynamic anarchies find home in colors, curls, and changing directions of dripped paint. The painting engulfs linguistic propositions which change direction and intersect, so that one cannot be sure which anarchy prevails when and how. Start, for example, from the upper-left of the painting. The orange curl seems to surface and cross with a white move which is not in a similar form as the orange curl. An orange arc next to it moves upwards and vanishes. The orange curl is superposed on black curls painted deeper. A white curl intersecting the orange one is partly marked by blue. Moving toward the center, we see that four blue specks are associated with white movements, and blue specks nowhere else mix except the white swirls in the whole painting. There is no orderly or successive movement in the painting, however. No cyclicality of colored moves prevents imagination of how the painting could extend in any direction and therefore "which have to be read like a written text" (Schapiro 1972–1973, 13). An assumption about the color of anarchies can be helpful here. Orange being a mixture of yellow and red, it can be interpreted as a security system of Hobbesian anarchy or as vitality and happiness, and therefore as a cooperative

Kantian anarchy. Regardless its color connotation the movement refers to albeit not complete but to a completed cycle. It is as if the anarchy the color connotes, either as a cooperative or a conflictive one, never transforms into another anarchy. Hence, if it is interpreted as a Kantian anarchy, it remains Kantian; otherwise, if it is interpreted as Hobbesian, it remains Hobbesian. However, there is change in both interpretations.

White movements do not cycle but twist without a power of the dominance of the black in the background of the painting. Naturally white has its own interpretation like innocence bridal gowns are argued to signify or surrender of an army in a war. Chinese wear white clothes in funerals, hence it can signify mourning after death. Hence, similar to orange cycle, white twists can be interpreted in different ways. It could be interpreted, for example, as disconnected security systems of acquiescence or consent among states. Does such an anarchy exist in Wendt's conceptualization? The answer is no. Perhaps one can argue that it depicts a cooperative system close but never equivalent to a Kantian anarchy.

Therefore, the form, color, and place of swirls, speck, and curls generate an interpretive mental engine to enrich and interpret Wendt's conceptualization of dynamic anarchies. These interpretations can be regulated in a disciplined fashion when they are thought to correspond to Peirce's triadic system of signs constituted by icon, index, and symbol. Interpretations of swirls, specks, or curls constituting an icon imply that these moves in the painting correspond meanings interpreters' minds form; they have semantic values. If swirls, specks, or curls constitute an icon of Wendt's dynamic and multiple anarchies, then by definition, they resemble Wendt's conceptualization. The orange and white movements in Pollock, if they are to be interpreted as icons, then they must imitate definitions and evolutions of the security systems they signify. Wendt describes these anarchies; he does not discuss how they evolve. The paint dripped on the canvas through his bodily movements signifies, in general, Wendt's assertion of anarchies as formed by states' practices constituting a dynamic process. The movement of paint and states' practices paving the way to alternative anarchies mutually correspond. They form an iconicity relation. Therefore, the curls, shapes, and arcs in Pollock's drip-painted works constitute icons of states' practices generating the three types of anarchies only, not the directions these anarchies take. The directions surface by further sign explorations.

The signs Pollock paintings contribute to Constructivism under the appreciation of empirical support being no guarantee for theoretical success. An agreement among IR scholars in sign making indicates how IR scholars accept which problems are important, how to approach these problems, and how to reach solution of these problems. The agreement attains and describes hidden realities about dynamic anarchies.

We note that Green and Tangerine on Red, The Number 14, Convergence, and Blue Poles lack perspective in the classical sense. Thus, the abstraction level of the inexistence of perspective of these paintings denotes a commonality across artworks mastered by different artists. The question then becomes whether it is impossible to construct signs for Structural Realism using Pollock's drip painting artwork and for Constructivism using Rothko's field paintings. The immediacy of producing interpretations for the respective artworks gazing these artworks helps to answer the question. Wendt's dynamic anarchies quickly find home in Pollock drip painting artworks unlike Waltz's structural constraints which become interpretable through the similarity of some Rothko works and Waltz's figure. The relation and correspondence of metaphoricity (Schwarz 1997) guides selections of the artwork. There exists a "figurative B-term" (paintings) and a "literal A term" (IR theory sentences) in a metaphor such that there exists a mental transfer from B-term to A-term (Forceville 1994, 1). Consequently, metaphoricity relation underlies interpretations of the artwork in Saussurean and Peircean terms, both. Interpretations of Saussure's signifier-signified and Peirce's icon, index, and symbol signs are all shaped by a correspondence between what one sees in material form that are paintings and what one attaches as meanings to those paintings roughly corresponding to a matter of thought, that is, mental interpretations, that can be qualified by Saussure's and Peirce's sign definitions.

3.4   Why Pollock and Rothko?

An IR theorist might prefer a masterpiece by Leonardo Da Vinci or Vincent Van Gogh instead of those selected in this book as more appropriate or meaningful or both for Structural Realism and Constructivism. One can take Picasso's "Guernica" to interpret, say, Structural Realism. There is no problem with such an artistic choice. Yet not anything goes when one interprets a painting to arrive at meanings of IR theories. A mimetic correspondence between ideas and abstract paintings facilitates the use of semiotic guides of Peirce and Saussure. The correspondence binds process of thoughts and strategic choices with those of visual examination. The correspondence stems from mimetic connections between Waltz's figure and Rothko's color-pane canvases and those between Pollock's drip paintings and Wendt's dynamic nonunique anarchies. Wendt's nonunique dynamic anarchies versus drip-paintings by Pollock pointing to the direction of repetitions deviate from structural constraints versus Rothko paintings generating color interaction feelings and sensations. Naturally, it is senseless to assert that Mark Rothko imitated Waltz's figure or that Pollock imitated Wendt's continuously changing dynamic anarchies. The correspondence allowing players to make meanings out of Rothko and Pollock's artworks for IR theories derives from these artworks' esthetic quality of facilitating perceptions and therefore interpretations through mimesis.

Thus, not all artworks are useful to interpret Waltz's and Wendt's conjectures and postulate signs. It is, for example, easier to interpret portraits, auto portraits, landscape, and still life paintings comparing them with, for example, Dutch School paintings of the same or a later era (Riegl 1999). Interpretive axis then would be evaluated as less arbitrary, and interpretations become easier. In comparison to such an art-art axis, the location of VSGs is to be found not at an artistic level but an ideational one at an Art-IR Theory axis. Paintings are material. Players see paintings. Dynamic anarchies and structural constraints are not observed or seen but they are mental. They are expressed in a linguistic form. The axis then transforms into a locus of idea-art as paintings constitute mental images of theoretical concepts proposed by Waltz and Wendt who do not paint IR, but they do so in ideational terms. Consequently, a nonarbitrary semiotic ground of sign making emerges by the help of the artwork of Pollock and Rothko.

The art of Rothko's color-pane Pollock's drip-paint works help to reach equilibria. Besides mimesis, selected paintings function as nonlinguistic systems (Goodman 1968, 226). It is not deniable that other paintings could help to construct signs. It is possible to explore further meanings of structural realist and constructivist propositions through alternative artworks. Neither Pollock nor Rothko paintings can be part of a meaningful communication among IR theorists unless the paintings constitute bases to interpret the two IR theories. But they do facilitate communication through mimesis and nonlinguistic systems they form.

Naturally, the states of mind of Pollock and Rothko are impossible to trace back. They can be momentary, individual, and unstable. Yet game theory and visual semiotics discipline states of mind of IR theorists. Nevertheless, communicating with each other, IR theorists have to probe about how their minds are disciplined along a semiotic axis. Even if players know mutual views about how artworks connect with IR theory statements, such knowledge runs short of how players' minds work. The book is centered around this contradiction between artists' and IR theorists' states of mind.

Chapter 4
Preferences and Equilibria

The elements of VSGs are the number of players, players' information conditions, players' strategies and actions, players' preferences over game outcomes, players' preferences over their actions leading to equilibrium or equilibria, and the specification of when the game ends. All VSGs are assumed to be two-player games of coordination. They model strategic interactions at two levels of analysis: extensive and strategic forms. Games in strategic form contain three elements: the number of players, players' strategies, and players' preferences over outcomes quantified by payoff functions. They are expressed by matrices. They abstract away players' information conditions by positing that players simultaneously choose strategies and receive payoffs. Simultaneous strategy choices correspond to the condition of imperfect information, that is, players ignore what strategy choice the other has made by others when they select a strategy. Coordination games in strategic form are extensively used in studies of human coordination (Schelling 1960; Gauthier 1975; Aumann 1976; Farrell 1987, 1988; Rabin 1990, 1991).

Games in extensive form are given by game trees. They are more complex compared to VSGs in strategic form. They specify the number of players, players' information conditions, actions available to players under different information conditions, sequences of players' actions, and the specification of when the game ends. Games in extensive form, also known as sequential-move games, model dynamics of actions through sequences of actions.[1] They allow for perfect information so that a player may be assumed to know what action the other player has chosen before she chooses hers.

---

[1] Repeated games where the same game in strategic form is finitely or infinitely repeated allow for perfect information of past interactions and descriptions of sequences of strategies.

© Springer Nature Switzerland AG 2023
S. Ş. Güner, *Art and IR Theory*, Mathematics in Mind,
https://doi.org/10.1007/978-3-031-32342-3_4

4.1 Preferences Over Outcomes

Preferences of players over outcomes of their interactions constitute the nuts and bolts of the engine that drives games. Explanations and justifications of preferences are necessary to apply game theory in strategic interactions meaningfully. Preferences should not come out of the blue. Yet there is a central difficulty in game-theoretic modeling: games are solved given players' preferences, that is, players' preferences are exogenous to game theory; they are assumed. Therefore, a fundamental aspect of VSGs remains outside the realm of game theory. Players' preferences are quantified by utility functions that remain outside the scope of game theory. Luce and Raiffa (1957, 12) assert that "utility theory is not part of game theory. It is true that it was created as a pillar for game theory, but it can stand apart and it has applicability in other contexts." Yet justifications and explanations of preferences are no easy matter: "Game theory is a tool for making decisions; but before you decide how to get what you want you must first decide what you want. Sorting out what you want−the function of utility theory−is not always as easy as it sounds" (Davis 1983, 57).

Game theory is a method to compute best replies in situations of strategic interdependence. Game theoretic aspect is players' preferences over their actions in Saussurean and Peircian games. Nash equilibrium and its refinements in a variety of interaction conditions expose players' preferences over their actions and constitute contributions game theory can provide in visual semiotics intersecting the Discipline. Nongame theoretic aspect of VSGs is preferences of players over outcomes of their interactions which game theory avoids dealing with. In the absence of players' preferences over outcomes no equilibrium can be computed. Hence, preferences over outcomes constitute the necessary condition for players' preferences over actions.

The neutrality of game theory about sources of preferences creates a wild range of possible hypothetical sources of preferences allowing to justify and explain players' preferences in VSGs. Hence, game theory applications in hypothetical interactions among IR scholars who discuss the meaning of paintings for IR propositions create a freedom space concerning preferences. Indeed, game theory is "completely neutral about what motivates people … tells you how to get what you want without asking why you want it" (Binmore 2007, 12). The diverse motivations of players shaping their preferences can be material or not: "The payoffs in a game needn't correspond to objective yardsticks like money or years spent in jail. They may also reflect a player's subjective states of mind" (Binmore 2007, 13). Players' subjective states of mind constitute sources of meanings, and, therefore, players' sign-making abilities. Therefore, players' preferences in VSGs reflect players' mental views about sign correspondences in Saussurean and Peircian terms. The problem of preference justification becomes critical especially for VSGs where players concur or not about signs they propose on the basis of aesthetic subjectivities. Each player proposes a sign gazing at a painting confirming Peirce's statement that "we think only in signs" (Peirce 1960a, b, c, 169). Hence, players' preferences stem as artifacts of their minds leading to the emergence of an interdependent semiotic language.

4.1.1 Preferences and Minds

Players' preferences over outcomes of their interaction represent phenomenological and psychological properties of players' minds (Chalmers 1996, 15). The role players' minds play in forming signs backed by preferences is summarized by the view that "our minds impose order on sensory data" (Breit 1984, 20). Thus, preferences reflect mental phenomena of players' sensations and propositional attitudes: "By sensations we shall mean bodily feelings like pains, tickles, nausea, as well as perceptual experiences like seeming to see a red pillar-box, hearing a loud trumpet, tasting a sweet strawberry" (McGinn 1983, 8). "The sensations and perceptual states" correspond to players' qualia (Audi 2015, 891). Players' sensations as qualia correspond to players' feelings like awe they feel when they are physically in front of a Rothko or a Pollock painting including their perceptual experiences like being immersed by the magnetism of the canvas. As to the propositional attitudes, they constitute beliefs triggering players' preferences over paintings. For example, a player's judgment such as "I believe that this painting by Pollock represents Wendt's nonunique dynamic anarchies" constitutes a propositional attitude.

Unlike propositional attitudes shaping players' intentions of choosing a specific strategy, qualia reflect players' sensations. Abstract art elements of colors, shapes, and their places on the canvas function as conveyors of feelings and therefore sensations and players' perceptive experiences. Qualia constitute bases of propositional attitudes like players' beliefs of a Pollock, or a Rothko painting constituting signs for Wendt's and Waltz's propositions, respectively. An IR scholar as a player can sense that a Rothko painting reflects color interactions, and a Pollock painting reflects on going dynamic anarchies and can select an action in a VSG accordingly. Therefore, propositional attitudes trigger strategy choices on the basis of qualia players' experience.

The connection between players' aesthetic subjectivities and games is given by Aumann (1985, 17): "If one thinks of mathematics as art, then one can think of pure mathematics as abstract art, like a Bach fugue or a Pollock canvas (though often even these express an emotion of some kind); whereas game theory and mathematical economics would be expressive art, like a cubist painting or Tolstoy's *War and Peace*." Hence, emotions as mental phenomena like sensations and propositional attitudes can help to explain players' preferences abductively in Peircian terms.

A coordination between players becomes possible if they reach a common code referring to the meanings of Pollock and Rothko paintings such as whirls and swirls in Pollock paintings and colors, places of colors, and color interactions in Rothko paintings through their subjective states of mind. For example, if we take Jackson Pollock as an example, he might be assumed to possess an inner need in letting paint drop on the ground to produce points, curls with thick and thin paints (Landau 1989, 168). A player might also have a similar inner need or urge to associate curls, points, changing directions of colors in a Pollock painting with changes of anarchies and state identities. The player would then see a Pollock painting as a representation of Wendt's anarchy conceptualization. Take, for example, Pollock's painting "Blue

Lines." An IR theorist associates those lines in the painting with different anarchies Wendt proposes (Sylvester 2001). Yet another theorist might associate them with different identities. Associations of paintings and theoretical elements are subjective.

Game theory does not exclude these subjectivities as nonsense. On the contrary, it takes them as central features of preferences driving strategic interactions. An immediate question then arises as what could the origins of these differences in associations? An answer would be the number of lines in the painting. As Wendt hypothesizes three anarchies, the number of lines would add another type of anarchy to Wendt's conceptualization. It might prompt a theorist to question the countability of anarchies as well. Perhaps one can propose four, five, or even a continuum of anarchies (Hopf 1998). The slopes of the lines similarly would help to distinguish differences among anarchies as one being steeper than the other so that it might be associated with an anarchy among friends or one among enemies.

The subjectivity of the association of painted lines with anarchies can be extended to associations between lines and states' identities. The problem then becomes while states can have identities of friend or foe, thus the number of identities being only two, the number of lines is four. In terms of iconicity, a Pollock drip painting containing two lines of different colors would be associated more "easily" with Wendt's identity conceptualization. There would then be a 1–1 correspondence between the hypothesized painting and the hypothesized number of state identities. Yet, more than two lines in a Pollock drip painting can be associated with more than two identities enlarging the space of "either friend or foe" dichotomy and yielding numerous identities. An added complexity is the color of lines and curls and their changing colors. The rhythm, color, direction, number of lines, and curls then must be assessed with respect to the rest of painting so that a player can interpret the painting as a tool to open new constructivist perspectives.

Players' subjectivities and strategy choices are distinct. The former belongs to the category of the mental while the latter to the category of objective game rules. Players' preferences are purely mental, yet they are assumed to be informed of strategies available to each of them, their preferences, and possible outcomes of interaction. A strategy could be players' confirmation that a painting constitutes a sign exposing an "internally consistent value system" (Schelling 1983, 4). Strategies connoting signs denote players' subjectivities including players' attraction to specific interaction outcomes. Thus, game rules and interactions are real; they are not mental constructions independent of players' minds. Players' minds alimenting preferences reflect how objective game rules based on subjectivities yield results of subjective sign-making processes.

Clearly, no objective connection between chosen paintings and IR theory propositions can exist. Signs are subjective. Any sign players agree upon does not exist in real terms; it cannot be "made true or false by facts in the world" (Blackburn 2016, 404). Players' interpretation of paintings as, for example, signifiers of some theoretical proposition is nothing but a human construction. Players' subjective tastes shaping their preferences constituting bases for real interactions imply a mixture of subjectivity and objectivity in all stability and instability of strategy pairs.

Making signs out of abstract paintings and making signs out of photographs like those of Abu Ghraib are similar as both sign-making processes require similar mental elaborations. The saying of "I don't know anything about art, but I know what I like" informs us that players must be IR scholars but not structural abstract art experts of Pollock or Rothko. They are assumed to be able to form their views and evaluations of a Pollock or Rothko painting as generating meanings for nonunique dynamic anarchies and structural constraints, respectively. The consistency of preferences allows to take players as rational interpreters. Therefore, there is need to discuss why players would agree or not on a common meaning of these paintings in a consistent manner.

Thus, the need for coordination derives from two sources: the social environment of the Discipline and players' mental instantiations and subjectivities like their sensations and propositional attitudes. Both players might evaluate paintings' meanings differently earning them unequal satisfactions players' minds instantiate. Game theory solves such complexities by displaying what signs strategically interdependent players select given their consistent preferences stemming from two sources. It clarifies how players' preferences underlie a failure or success in the creation of a semiotic language and reveals preferences over actions given preferences over outcomes.

I discuss two examples to introduce sources of preferences without entering formalities of equilibrium computations that come in the next chapter. The examples are prisoner's dilemma and the dress-color game developed from an idea of Perea (2012). They expose the central problem of the clarification of players' preferences in general and consistency of players' preferences.

4.2 Preferences Over Actions: Nash Equilibrium

Game theory does not deal with preferences over outcomes but it deals with players' preferences over strategy choices. A Nash equilibrium is a pair of strategies such that no player has any incentive deviate from her strategy given that the other player sticks to hers. All equilibrium strategies are best replies against each other and "satisfy players' incentives" (Aumann 1985, 19).

There are two players who select between two strategies in the generic 2×2 game in the strategic form given below. Row chooses between her strategies of up and down and Column chooses between left and right. Row obtains utilities a, c, e, and g, and Column b, d, f, and h in upper-left, upper-right, lower-right, and lower-left cells of the game matrix, respectively. Nash equilibrium conditions are as follows (Table 4.1):

(Up, Left) pair of strategies constitutes a Nash equilibrium if and only if $a \geq g$ and if $b \geq d$;

(Up, Right) pair of strategies constitutes a Nash equilibrium if and only if $c \geq e$ and if $d \geq b$;

Table 4.1 A Generic 2 × 2 game in strategic form

Row		Column	
		Left	*Right*
	Up	a, b	c, d
	Down	g, h	e, f

(Down, Left) pair of strategies constitutes a Nash equilibrium if and only if $g \geq a$ and if $h \geq f$;

(Down, Right) pair of strategies constitutes a Nash equilibrium if and only if $e \geq c$ and if $f \geq h$.

O'Neill (2001, 271) presents two approaches to justify why these strategy profiles constitute Nash equilibrium: either equilibrium conditions are abstract properties or players use "contextual knowledge": "Rationales for the Nash equilibrium turn on a certain premise that is often unspecified. Will the players be coming to their decisions only from the properties of the abstract game, or will they use other contextual knowledge available to them? The first approach has been called the *tabula rasa* assumption, and the latter could be called the *relevance of context*."[2] Our approach takes a middle position: players choose strategies by developing mental calculations once their preferences contain information about the context, that is, the Discipline in a constant turning mode and the need for coordination combined with esthetic judgments steered by semiotics. The abstract nature of the game contains information about the context in terms of players' preferences. Players' preferences differ across VSGs but the approach to detect stability in interactions remains the Nash equilibrium.

4.2.1 Prisoner's Dilemma

The game of Prisoner's Dilemma at strategic form given below opens a range of reflections and conjectures about the origins of preference orderings. Row and Column are the players.[3] They both prefer less prison time. Therefore, each player prefers the most to confess the crime they together have committed while the other remains silent. The next best outcome for each is the one where they remain silent together. The next-worst outcome for both is their joint confession. The least preferred, that is, the worst outcome, arises when one remains silent but the other confesses. The silent one gets a sentence of, say, several years of prison while the confessor is out in few weeks (Luce and Raiffa 1957, 95):

---

[2] Italics in the original.

[3] Row player is considered as male and Column player as female from hereon.

To obtain a game of Prisoner's Dilemma at ordinal level, outcomes are categorized and ranked indicating players' payoffs. Suppose that players try to minimize their prison time as they are assumed to be selfish and greedy. Now the game matrix emerges under the preference assumption covering both players.

If one confesses while the other does not, then the confessor obtains her best payoff and the one that remains silent obtains her worst payoff. If both remain silent, they obtain their next-best payoff. If both confess the crime, they obtain the next-worst payoff. The ranks of 4, 3, 2, and 1 indicate the best, the next-best, the next-worst, and the worst outcomes in terms of prison times for both players, respectively. Thus, the game at ordinal level becomes:

What are the optimal choices of the players? Will they remain silent or cooperate? We answer the question by solving the game for the best strategy profile. There exist four strategy profiles in these games: (remain silent, remain silent), (remain silent, confess), (confess, remain silent), and (confess, confess).

Each player knows the rules of the game comprising information conditions, strategies available to each player, the number of players, and players' preferences and each knows that the other knows the rules of the game and each knows the other knows that each knows the rules of the game, and so on ad infinitum. Hence, this is a game of complete information.

The game of prisoner's dilemma models the fear of being cheated. The fear felt stems from prisoners' subjective mental states. Both players have an incentive to confess because it is the best strategy choice regardless of what choice the other player makes. For example, from Row's perspective, if Column remains silent, confessing is better than remaining silent. It leads to an outcome of 3 months of prison instead of 1 year. If Column confesses, then again confessing is a better strategy choice because 8 years of prison is better than 10 years. Thus, the strategy of confess is always better than the strategy of remain silent. Column's position is symmetric compared to Row's, consequently, Column's strategy of confess is always better than her strategy of remaining silent. Therefore, the strategy of "confess" is called a "dominant strategy." Both players believe that each prefers to select confess and each believes that the other believes that the other will confess.

The strategy pair of (confess, confess) constitutes a dominant-strategy equilibrium, that is, the stable outcome of the game which no player would have an incentive to deviate from given the other's strategy selection. Players cannot reach the better outcome of (remain silent, remain silent) illustrating the dilemma Row and Column face. Consequently, the stable outcome reflects strategic stability due to players' motive to reduce their prison time. Alternative motives of suspects would create games different than the Prisoner's Dilemma.

Preferences are consistent with players' motives and states of mind. The immediate question becomes this: why players' states of mind is fixed on terms of prison years or months? Are there no other criteria than time passed in prison to measure and generate players' preferences over the four outcomes? What preference transformation would occur, for example, if players are so close friends that one or both prefer to remain silent on purpose because they value being silent as an honorable action? If this is the case, then Tables 4.2 and 4.3 does not reflect players' true

Table 4.2 Prisoner's dilemma outcome matrix

Row		Column	
		Remain silent	*Confess*
	Remain silent	1 year each	10 years for row and 3 months for column
	Confess	3 months for row and 10 years for column	8 years each

Table 4.3 Prisoner's dilemma game

Row		Column	
		Remain silent	*Confess*
	Remain silent	3,3	1, 4
	Confess	4, 1	2,2

preferences, and, accordingly, players are not involved in a Prisoner's Dilemma. Hence, the prisoners in custody interact in a different game in which the strategy of confess is unethical. The social norm both or one of the prisoners adhere to is keep silent about a crime. It is possible that players' motives are "socially constructed and deeply dependent upon cultural experience and social interaction...most humans are inextricably *social creatures* whose preferences are affected by moral considerations that are situationally specific. In particular, people tend to conform to *social norms* that are reflected in the priorities represented in their preference orderings" (Gintis 2000, 31).[4] The level of friendship between the players or the social environment players belong to can eliminate circular feelings of the fear of being cheated. Three months compared to 1 year of prison is a shorter prison time but a heavy burden on the shoulders of the confessor under players' friendship or social environment. Prison terms can be unchanged but players' evaluations of them can differ. Players can prefer longer prison times under different motives like being close friends or simply fearing that cheating will be punished by, say, the mob members outside. Hence, it is only one of the criteria to evaluate outcomes in sheer meaning of prison time; there are alternative sources of preferences than selfishness and greed: "It is a common mistake (unfortunately one that even game theorists make) to reason under the assumption that players are selfish and greedy. This is, typically, an unwarranted assumption. Research in experimental psychology, philosophy and economics has amply demonstrated that many people are strongly motivated by considerations of fairness" (Bonanno 2015, 8).

If scholars interact once, the alignment of preferences does not preclude the possibility of players' obtention of unequal utilities. Thus, while Row and Column's minds meet on the iconicity of a Rothko painting for structural constraints, they

---

[4] Italics in the original.

Table 4.4 VSG as Prisoner's dilemma at ordinal level

Row		Column	
		Accept	Reject
	Accept	3,3	1, 4
	Reject	4, 1	2,2

value their meeting of minds unequally. It is possible to model a VSG where Row and Column's preferences partly coincide on the outcomes of their interaction such as one prefers Rothko as a signifier, but Column does not. A prisoner's dilemma as a VSG would represent how coordination fails as both players find that the Rothko painting is not a signifier for structural constraints. The VSG below takes the form of prisoner's dilemma where accept and reject correspond to "Rothko is a signifier for structural constraints" and "Rothko is not a signifier for structural constraints," respectively (Table 4.4).

The game's focus now turns on the fear of IR scholars being cheated. While if both accept they reach agreement but they cannot accept, because each fear that "if I accept, the other can reject, and I obtain my worst outcome" and "If I reject, I reach higher utilities regardless what the other does." The game then would expose a reasoning of the failure to coordinate in the profile of "accept, accept" and the success in the profile of "reject, reject" as the unique stable outcome. No coordination in the profile of "accept, accept" can then ever be possible in the VSG taking the form of prisoner's dilemma. Nevertheless, it is possible to demonstrate that "accept, accept" outcome emerges as a stable equilibrium provided that the game of prisoner's dilemma is modeled as an infinitely repeated game and scholars are sufficiently patient with respect to the future and avoiding short-term cheating (references here). The repetition of the same prisoner's dilemma game is not the only way to model coordination. Imagine that scholars interact in a two stage-game where the first stage is the prisoner's dilemma, and the last stage is a pure game of coordination like the battle of sexes. Assuming that the first stage equilibrium is (reject, reject) and scholars interact in the battle of sexes game either by certainty or by some probability at the second stage, it is possible to study how coordination can emerge under similar mixed contingencies.

Overall, Prisoner's Dilemma posits players' utilities lie in the length of prison time; however, prison time is not the unique criterion for the players to form consistent preferences. Alternative sources of preferences over the same interaction outcomes yield alternative games. The principal problem of applying game theory in empirical realm is exactly the clarification of unobservables prompting players to rank and order outcomes of their interactions. The Dress-Color game permits more reflections on this problem.

4.2.2   Dress-Color Game

Perea (2012: 17) offers a decision problem (not a game) without going into the sources of preferences: "This evening you are going to a party with your friend Barbara. The big problem is: Which color should you wear tonight? Suppose that in your cupboard you have blue, green, red, and yellow suits, and the same holds for Barbara. You prefer blue to green, green to red and red to yellow. However, the situation you dislike most of all is when Barbara wears the same color as you!" It is instructive to transpose this one-person decision-making problem into a game. Suppose Barbara, Alice's friend, is invited to the same party. Which color choice should Alice make? Both friends engage in a conversation about colors of dresses they would wear in the party. Barbara and Alice now interact and therefore they are involved in a game as one's decision depends on the other's decision. The interaction forces one to explore and guess sources of players' preferences producing a multitude of alternative games.

Imagine that both Alice and Barbara have identical preferences: both prefer blue to green, green to red, and red to yellow but try to avoid wearing a dress of the same color. The critical question is why do Alice and Barbara have such preference orderings? Would not it be that Barbara has different color preferences, but she does not hate wearing a dress of the same color as Alice does? Or would not be that Barbara has the same color preferences as Alice, but she avoids wearing a dress of the same color Alice selects? What are the factors shaping Alice's dislike to wear a dress of the same color as her friend Barbara's, and, distinct from that, her preferences over dress colors? These questions pertain to subjective tastes of Barbara and Alice.

Alice's (likewise Barbara's) preferences over dress colors are difficult to explain, they eventually stem from her subjective tastes strictly related to the color wheel. Yet her urge of not to wear a dress of the same color as Barbara is open to alternative explanatory hypotheses through abduction (Peirce 1955). Abduction constitutes a type of explanatory inference. It is different from deduction that searches for the implication of the truth of a conclusion from the truth of the premises, like, say, modus ponens or modus tollens (Copi 1961). It is also different from induction claiming the obtention of a universal result from a finite number of observations by, for example, using statistical tools. In abductive terms, there can be different motives explaining factors that push Alice and Barbara to avoid wearing dresses of the same dress color. For example, it may stem from Alice's craving or compulsion to be noticed in the party or Alice's psychological make up might attribute a high importance to what others think about her in the party. Naturally, there is a range of other factors helping both players in the party, yet simplification needs push to select dress color in the interaction. Thus, Alice's urge to avoid wearing a dress of the same color as Barbara does springs from her social environment encouraging uniqueness and her subjective tastes. Perhaps, the dominant social norm might be to wear a dress with a unique color with no match with those of the others. Alice might therefore attribute a high priority to be in conformity with those norms. Wearing a dress with a unique color, Alice experiences a certain sensation of getting attention of the partygoers regardless of their sex. The experience then generates feelings of happiness in Alice's mind. In other words, Alice lives through a "conscious experience" in the party (Chalmers 1996, 3). A clue to assess sources of preferences of Barbara and Alice is offered by Wildavsky (1987, 3) who indicates that "preferences come from the most ubiquitous human activity: living with other people. Support for and opposition to diverse ways of life, the shared values legitimating social relations (here called cultures) are the generators of diverse preferences." Therefore, Alice's and Barbara's urge to be noticed and their preference to avoid wearing a dress of the same color constitutes a way of life, a result of living with other people, and a sign of conformity with a culture of parties. As to their preferences over colors, they are of personal nature. Alice and Barbara's preferences over colors reflect their color tastes, that is, "subjective states of mind" (Binmore 2007, 13). Players' subjective states of mind reflect players' satisfaction in outcomes of their interaction. "A useful way of thinking of utility is as an 'index of satisfaction'" (Bonanno 2015, 11).

The whole sequence of arguments explaining dress-color preferences of Alice and Barbara constitutes an example of abduction corresponding to the concept of the inference to the best explanation (Harman 1965). Social norms and the motive of being noticed in the party provide strong reasons to believe in the correctness of the sources of preferences of Alice and Barbara (Lipton 2000). It must be noted that the motive of being noticed in a party does not rigorously imply preferences of Alice and Barbara. The motive of being noticed in a party is nothing but a member of a set of alternative explanatory hypotheses. Yet it might the best explanation because it is simple and perhaps verifiable by observations and data.

4.2.2.1 Nash Equilibrium in Dress-Color Game

Perea proposes the following preference ordering for Alice: Alice prefers a blue dress to a green one, a green dress to a red one, and a red dress to a yellow one. Blue is assigned a value of four, green a value of three, red a value of two, yellow a value of one, and the value of wearing a dress of the same color as Barbara is zero. Suppose that Barbara's color preferences are identical to those of Alice. The situation then easily evolves into a strategic interaction and therefore to a game if Barbara and Alice must think about each other's dress color choices. Barbara must think about Alice's color choice and Alice must think about Barbara's color choice.

Each might prefer to be in conformity with the social norm. Each might be conscious of the fact that omitting the other's choices cannot be realistic in this "game of getting (or competition for) others' attention in the party." If both choose the same color, they will not experience those mental feelings of satisfaction and being in conformity with the party culture. Consequently, a game in strategic form can be proposed where Alice and Barbara are the players, blue, green, red, and yellow are strategies, and preferences are as given above. The game matrix is given below (Table 4.5).

Table 4.5 Dress-color game

Alice	Barbara			
		Blue	Green	Red	Yellow
	Blue	0, 0	4, 3	4, 2	4, 1
	Green	3, 4	0, 0	3, 2	3, 1
	Red	2, 4	2, 3	0, 0	2, 1
	Yellow	1, 4	1, 3	1, 2	0, 0

There are 16 outcomes corresponding to categories and ranking of these categories in the game. The first number in each cell corresponds to the payoff of Alice and
the second number corresponds to the payoff of Barbara. For example, if Alice
wears green and Barbara blue, Alice obtains a payoff of 3 and Barbara a payoff of
4. The diagonal of the game matrix consists of zero values as Alice and Barbara
wear dresses of the same color. The rest of the cells correspond to Alice's and
Barbara's personal color tastes and therefore their subjective states of mind.

The game is at the ordinal level of measurement because players' preferences do
not expose the strength of preferences. For example, there is no information about
how much Alice or Barbara prefers the outcome where Alice wears red and Barbara
wears green over the outcome of Alice wears red and Barbara wears blue. Similar to
Alice, Barbara prefers blue over green but there is no information like Barbara
adores blue so much that Barbara highly despises green next to blue. Ordinal preferences are about each player's own subjective satisfaction from the outcomes. They
give no information about the intensity of preferences.

Alice and Barbara are involved in a 4 × 4 game in strategic form. They decide on
their dress colors without knowing each other's choices under imperfect information. They do not see each other in dresses they will put on for the party before they
pick their dress color. Otherwise, if one of them picks a color and the other moves
by selecting a color, then a dress-color game in extensive form becomes the correct
level of analysis. Neither Barbara or Alice has a dominant strategy like in the
Prisoner's Dilemma game. Thus, it is impossible to solve the game through dominant strategies. The concept of Nash equilibrium must be used to solve the game.

In a Nash equilibrium, players cannot improve their payoffs by changing strategies given the strategy choice of the other in the game. Consequently, Nash equilibrium in the dress-color game is a pair of strategies such that neither Alice nor
Barbara would have any incentive to deviate from her strategy given that the other
sticks to her strategy. If we look at this game from Alice's point of view, we notice
that Alice has an incentive to choose blue unless Barbara chooses blue. The choice
of blue can lead to a payoff of zero if Barbara chooses blue; otherwise, if Barbara
chooses colors other than blue, then blue is the best choice. Barbara is in a symmetric position. Both have an incentive to avoid choosing the same color. If Alice
chooses blue then she obtains her most preferred dress color blue, therefore the
payoff of 4, given that Barbara chooses green. Barbara obtains a payoff of 3 by
replying with green against Alice's choice of blue. Given Alice's choice of blue,
blue gives a payoff of 0, green 3, red 2, and yellow gives a payoff of 1 for Barbara.

Therefore, Barbara's best response to blue is green. Given Barbara's choice of green, blue gives a payoff of 4, green gives a payoff of 0, red gives 2 and yellow gives 1 for Alice. Therefore, Alice's strategy of blue against Barbara's strategy of green and Barbara's strategy of green against Alice's strategy of blue constitute best replies forming a Nash equilibrium of the game. Alice choosing green and Barbara choosing blue constitutes another equilibrium as these choices are best replies against each other as well. Hence, either Alice or Barbara will wear a blue dress. The equilibrium does not identify who will wear the blue dress in the party.

There is no other equilibrium in the game. Take for example, the pair of Alice chooses green, and Barbara chooses red. If Alice chooses green, red gives a payoff of 2 to Barbara. If Barbara deviates to blue, then she obtains 4, her best utility in the game. Barbara would then deviate to blue given that Alice chooses green. Consequently, there are only two equilibria in the game: Alice wears blue, and Barbara wears green and vice versa. The game implies that it is impossible to see Alice and Barbara wearing dresses of the same color. The two equilibria inform how players' subjective color tastes and inner lives feeding their self-esteem in being noticed in the party under cultural constraints drive their strategy choices. They illustrate how the culture of the community Alice and Barbara live in constrain color choices of Alice and Barbara, how they get mental sensations of getting attention of the partygoers resulting in their feelings of happiness, and how their personal color tastes affect their preferences over actions leading to stability.

Alice and Barbara form beliefs about others' beliefs shaping the equilibria in the game. Alice expects Barbara would never choose a red or a yellow dress. Barbara similarly expects Alice would never choose a red or a yellow dress. Each believes that the other would choose green against blue and blue against green. Alice knows that Barbara knows Alice knows the best reply against green is blue and vice versa. Barbara is in a symmetric position. Players' positions derive from the common knowledge condition: Barbara and Alice know the game rules, each knows that that the other knows, the other knows that each knows the other knows…ad infinitum the rules of the game. Each would avoid wearing a blue dress if each believes that the other wears a blue dress. If each wears a green dress due to their beliefs that they wears blue, then the outcome will fail to satisfy Nash equilibrium conditions. Hence, the equilibrium depends on what players believe about beliefs of the other.

4.2.2.2 Dress-Color Game Solution by Backward Induction

An immediate question arises as whether the equilibria change if Alice and Barbara are informed reciprocal color choices. We have to offer the dress-color game in extensive form to answer the question. Games in extensive form are a bit more detailed than strategic form games. They specify players' information conditions, actions available to players in these information conditions, sequence of players' actions, and players' preferences over the outcomes are described, and when the game ends. Hence, in the dress-color game in extensive form we must specify who

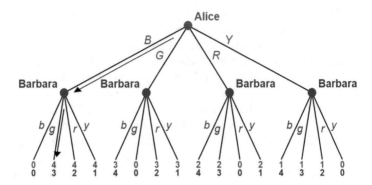

Fig. 4.1 Dress-color game in extensive form. (The game tree is produced using the software developed by Rahul Savani and Bernhard von Stengel (2015), "Game Theory Explorer – Software for the Applied Game Theorist." *Computational Management Science* 12, 5–33.)

starts the game, who moves next under which information conditions, and when the game ends. The game tree is given below (Fig. 4.1):

The initial node is where the game starts. Suppose that Alice moves first by selecting a color among blue (B), green (G), red (R), and yellow (Y). She has perfect information as she knows that she starts the game. Barbara selects a color among four after Alice by observing Alice's choice among four colors. Therefore, Barbara is said to have perfect information when she reacts to Alice's color choices. Barbara reacts to each of Alice's color choice by choosing among blue (b), green (g), red (r), and yellow (y) dresses. The game ends when Barbara selects a color.

Suppose, for example, that Alice starts the game by selecting a blue dress. Barbara now must decide which color she will wear being informed of Alice's action. A scenario would be that Alice invites Barbara over her house and shows Barbara that she will wear a blue dress before the party starts. Barbara can react by choosing blue as well, but if she chooses blue, both will obtain zero payoffs. Hence, Alice's invitation could work against Barbara's interest in showing off at the party. Had Alice chosen any other color than blue, then blue becomes Barbara's optimal color. Barbara picks green, her next-best alternative, to avoid the prospect of wearing blue dress facing Alice's selection of blue. Any reaction to blue by selecting red or yellow generates lower payoffs than green. Therefore, her reaction will be green against Alice's choice blue.

The solution of Alice wears a blue dress and Barbara a green one is found by the method of backward induction based on the principle of "look ahead and reason back" (Dixit and Nalebuff 1991, 34). The solution identifies the sequence of actions working by first finding out Barbara's optimal choices while she moves the last in the game: if Alice selects blue, she responds by selecting green. If Alice selects any other color than blue, then she responds by blue. Alice anticipates Barbara's reactions to her choices. She does not select green, red, or yellow but blue to force Barbara to choose green in order to obtain her best outcome in the game. She foresees that her choice of green leads to the outcome where Barbara wears blue, and, therefore, she obtains a payoff of 3, and, if she chooses red or yellow, Barbara wears again a blue dress and she obtains 2 or 1, respectively. Thus, Alice's choices of green, red, and yellow lead to Barbara wearing blue. But Alice prefers blue at most. If she selects blue at the start, Barbara is left with nothing but a choice green. Consequently, the equilibrium of the game is composed of Alice's choice of blue and Barbara's choice of green. The solution eliminates the equilibrium the game in strategic form generates where Alice wears a green dress and Barbara a blue one. If Alice moves first, she reaps the advantage of shaping the equilibrium in her favor (Schelling 1960, 48). Similarly, if Barbara moves first, then Alice will be forced to choose a green dress. Therefore, a change in information conditions alters the equilibrium in the game by privileging the first mover; the player who moves first is able to make the second mover to select green. In short, if Alice or Barbara moves first by selecting a dress color, then she has power over the other by forcing the other to choose green.

4.3   Coordination

How does game theory shed light on constantly turning IR discipline? The answer I propose is through examination of coordination games that are Pure Coordination, Stag Hunt, and Battle of Sexes games. There are two Nash equilibria in all 2 × 2 coordination games modeling interactions between two players each possessing two strategies. Each game exposes intricacies and differences of coordination problems among players who are IR scholars. Each generates solutions by their equilibria that derive from payoff configurations reflecting players' preferences.

In a pure Coordination Game, players obtain equal payoffs in equilibrium outcomes. In the Battle of Sexes, the Nash equilibria favor one player unlike the other yet discoordination is painful for both. In Stag Hunt games Nash equilibria can generate equal or unequal payoffs yet a unilateral selection of defection, while not generating an equilibrium, can be more beneficial than what the player obtains in an equilibrium. Thus, varieties of coordination games describe alternative processes of players' convergence in the sign quality of the pair of art and IR proposition.

Why coordination games? The answer requires players' interest in coordination and their evaluations of outcomes of their strategy choices. Players prefer coordination as world is rapidly changing vying for explanations, concerted or not. The other reason is that IR scholars perceive the Discipline in turmoil (Lapid 1989; Wæver 1996). They value theoretical communication, either linguistic or not, as vital for the Discipline. They do not exclude any communication using abstract art. It is not impossible or senseless for them to interpret structural constraints and nonunique dynamic anarchies by the help of art of Mark Rothko and Jackson Pollock. A question could be "why abstract art?" An answer could be "why not?" If the Discipline does not exclude evolutionary biology, leaders' gestures, what garments world leaders wear, photographs taken at sites of crisis, wars, and human rights abuses, then there is no place for knee-jerk rejection of abstract art in theoretical communications.

Players might have seen no Pollock or Rothko painting in their lives, they might ignore them, or dislike abstract art altogether. But none of these factors would prevent them to possess cognitive processes enabling mental connections between art and IR propositions in linguistic form. Zajonc (1980, 160) states that:

> The prevalent approach to the study of preferences and related affective phenomena holds that affective reactions follow a prior cognitive process: *Before I can like something I must first know what it is.* According to this prevalent view, therefore, such cold cognitive processes as recognition or categorization are primary in aesthetic judgments, in attitudes, in impression formation, and in decision making: They come first. If we say, for example, that we like John because he is intelligent, rich, and compassionate, it follows that we must have gained some impression of John's intelligence, wealth, and compassion, and combined them, before we formed an attraction to him."[5]

Zajonc's argument implies that VSG players first see a painting by Pollock or Rothko and only after that they ponder about whether and how the painting can generate a specific meaning for the proposition by Wendt or Waltz. Thus, there are two steps in players' preference formation: first players see a painting and decide whether the painting is helpful in constructing a meaning. They form later a mental correspondence between the painting and the proposition forming a preference ordering over interaction outcomes. Players select strategies on the basis of a cognitive approach weighing pros and cons of their choices. They can either choose that the painting is a signifier, an icon, index, or a symbol based on their calculations. Consequently, equilibria of VSGs reflect shared meanings through strategic interactions and open a new communication channel in semiotic and game theoretic terms.

Myerson (1991, 250) indicates that "a game is with communication if, in addition to the strategy options explicitly specified in the structure of the game, the players have a very wide range of implicit options to communicate with each other." VSGs as games of coordination help to study how a common semiotic language emerges allowing a fruitful theoretical communication through players' minds exposing conditions for the emergence of common theoretical perspectives. Players who do not exclude art as a basis of interpretation of IR theory sentences should also have preferences over sign quality of paintings like whether they are signifiers or the painting in question stands as an icon, index, or symbol of an IR theory sentence. Players of VSGs do not explain an observed IR event but interpret abstract IR propositions. As they interact in a coordination game, they are conscious of their strategic uncertainty in proposing signs.

Players are aware that a unilateral defection corresponds to missing an opportunity of arriving at a mutual understanding, a convergent interpretation. They act on reasons deriving their beliefs about the Discipline in turmoil. It would of course constitute a category mistake (Ryle 1949) to equate a painting with an IR proposition; there exists no equivalence relation between the two. For example, players' feelings emanating from the artwork and players' revisions of their beliefs about an IR theory on the basis of these feelings and locating the source of these feelings in their brains require a proof. How can these subjective and conscious experiences be

---

[5] Italics mine.

brain products? The question reveals the role of alternative views of the philosophy of mind in VSGs. A coordination game implies that each player knows the preferences of the other and that each knows that the other is under the same psychological pressure of avoiding discoordination and each knows the other knows that the other faces incentives to avoid coordination failures and so on ad infinitum exposes the complexity of interdependent preferences and minds in any VSG.

Players interpret the artwork to arrive at an undiscovered meaning of the proposition contributing to the Discipline. The semiotic language indicates a commonality in plastic thoughts facing a painting Pollock or Rothko. Common interdependent interpretations of these paintings for IR theoretical statements corresponding to signs indicate scholarly agreements on the meaning of propositions through art and game theory. Consequently, paintings in question leave the domain of art and enter the domain of the Discipline through their abstract plastic meanings representative of an IR theory statement. Signs being coordination products delimited by languages of semiotics and game theory offer precise concepts, relations among concepts, and suppositions. Semiotics and game theory together delimit strategic interactions and discipline subjective interpretations. Not anything goes in interpretations and communication of these interpretations. There exists a strategic interpretative order VSGs impose on scholars' communication and language they use in their interactions.

Coordination is modeled by VSGs in which both Row and Column can concur that a painting constitutes (or not) a sign. Hence, players' preferences can be aligned. The question is why would Row and Column have aligned preferences? Schelling's answer is instructive:

> It may be important to emphasize that, in referring to a "common interest," I do not mean that they must have what is usually referred to as a similarity in their value systems. They may just be in the same boat together: they may even be there only because one of them perceived it a strategic advantage to get in that position—to couple their interests in not tipping the boat. If being overturned together in the same boat is a potential outcome, given the array of alternatives available to both parties, they have a "common interest" in the sense intended in the text. "Potential common interest" might seem more descriptive. Deterrence, for example, is concerned with coupling one's own course of action in a way that exploits that potential common interest" (Schelling 1983, 12).

IR scholars who are players in VSGs are in the same boat in Schelling's terms. They are members of a turbulent Discipline continuously turning around and evolving through debates. Communication problems stemming from meta-theoretical controversies and maelstroms of a continuous introduction of new theories find an alternative research exit in a synthesis of semiotics and game theory.

Social environment creates a powerful incentive to be in the same boat. IR scholars publish, make research, offer theories in meeting and communicating with each other, in short, they form a common social environment corresponding to the sources of preferences is "living with other people" according to Wildavsky (1987, 3). Players' VSGs are under similar social environment constraints. Players prefer coordination but not all coordination instances might benefit them equally. For example, in the dress-color game, Alice and Barbara have their subjective preferences over the color wheel leading to alternative utilities of equilibrium. In VSGs, these subjectivities cannot be defined in a basic mode, because, for example, seeing a drip-painting canvas by Pollock and associating it with Wendt's conceptualization of nonunique dynamic anarchies is an example of instantiation; the painting instantiates in the mind of a player the property of dynamic nonunique anarchies. Instantiation, as a concept offered by McGinn (2012), supports the idea of when, for example, Row and Column see a canvas by Pollock, they each perceive it as representing or not dynamic nonunique anarchies so that the painting is a Saussurean signifier or not or a Peircian sign like icon, index, and symbol. They might not have an identical instantiation, yet they can face incentives to coordinate even though they do not value coordination equally.

4.3.1 Pure Coordination

The game of pure coordination is the fundamental model of the coordination problem (Table 4.6).

The pure coordination game implies that players have nothing but choosing same strategies by reciprocity. The Nash equilibrium of CC and DD gives both players equal satisfaction. Either cooperate together or reject together are equivalent in terms of players' utilities. The game does not however specify which equilibrium will be selected out of the two. Suppose that Column selects C with a probability of p and therefore D with a probability of $1 - p$. Hence, Row obtains an expected payoff of p (a) + $(1 - p)$ (0) from the choice of C and p (0) + $(1 - p)$ (a) from the choice of D. Row would be indifferent between choosing C and D as her expected payoffs to C and D are equal, therefore if p (a) + $(1 - p)$ (0) = p (0) + $(1 - p)$ (a), that is, if $p = \frac{1}{2}$. Consequently, Row will choose C as long as she believes that Column's choice of C is greater than $\frac{1}{2}$. Column is in a symmetric position; therefore, Column will choose C if she believes that Row will choose C with a probability higher than $\frac{1}{2}$ (Lewis 1969, 25). If players communicate prior to their strategy choices and agree to play C with probabilities higher than $\frac{1}{2}$, then the coordination on C is reached; otherwise, if players communicate prior to their strategy choices and agree to play D with probabilities higher than $\frac{1}{2}$, then the coordination on D is reached. The probability of a successful coordination changes with coordination failures becoming increasingly costly. For example, if the upper-right cell of the game above where Row chooses C and Column chooses D contains high costs for both players,

Table 4.6 The pure coordination game in strategic form

Row		Column	
		C	D
	C	a, a	0, 0
	D	0, 0	a, a

say, $-c$, then the game does not still specify which equilibrium will be selected but the magnitude of p, that is, Column's probability of selection of C, will change. Row obtains an expected payoff of p (a) + $(1 - p)$ $(-c)$ from the choice of C and p (0) + $(1 - p)$ (a) from the choice of D. Hence, Row will choose C if p (a) + $(1 - p)$ $(-c) > p$ (0) + $(1 - p)$ (a); if $p > (a + c)/(2a + c)$ and will choose D if $p < (a + c)/(2a + c)$. We observe that Row selects C only if she attaches a higher likelihood to Column's choice of C because Row's choice of C countered by Column's choice of D is costly. Row wants to be sufficiently certain that Column reciprocates her choice of C; otherwise, she is inclined to select D to reach coordination at DD outcome. Players can communicate before they make their choices, but they cannot be certain that no mistakes are run when they interact actually. Players can obtain different payoffs in coordination outcomes reflecting a conflict of interest as the Stag Hunt and Battle of Sexes games illustrate below.

The pure coordination game in extensive form below clarifies that the uncertainty of the equilibrium selection does not dissipate if the game is represented in extensive form (Fig. 4.2):

Column reciprocates Row's choices she is informed of. Row obtains equal utilities by choosing C and D; she has no strict preference between them. Both CC and DD profiles constitute Nash equilibria; no unique equilibrium ensues players' sequential interaction. Therefore, either sequential or simultaneous, players' actions yield two stability outcomes. Players' coordination reflects players' indifference with respect to ways to coordinate actions. Sequential choices eliminate one of the equilibria in Stag Hunt and Battle of Sexes games as shown below.

4.3.2 Stag Hunt

The original story of the game of Stag Hunt (also known as the Assurance Game) is that a group of people, say a tribe, is after hunting a stag. If they all participate, they will succeed in hunting the stag. Yet each member can leave the joint hunt if he sees a rabbit, catches, and eats the rabbit to satisfy his hunger. If one follows the rabbit,

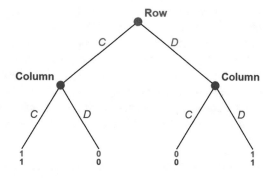

Fig. 4.2 The perfect information pure coordination game in extensive form

Table 4.7 Stag Hunt

Row		Column	
		C	D
C		4, 4	1, 3
D		3, 1	2, 2

then the whole hunt for the stag fails. Naturally, it is better to hunt the stag to eat more than eating the rabbit but incentives of leaving the hunt are clear and present as hunters are hungry.

The game matrix at ordinal level is given below (Table 4.7):

The strategy C denotes cooperation, that is, involvement in the hunt for a stag. The strategy D denotes defection, that is, leaving the joint hunt and chasing a rabbit instead.

The problem Stag Hunt game underlies the following problem: while mutual cooperation represents the best outcome for both players, it is not a guaranteed one. There is a risk for the stag hunter: if the other hunter sees a rabbit and runs after it, ruins his dream of eating together large chunks of meat. Players' joint preference to hunt a stag as a better hunt compared to rabbit does not constitute any sufficient condition yielding the best outcome for both players. Therefore, the game of Stag Hunt models the possibility of a productive coordination through mutual cooperation and a nonproductive coordination at mutual defection, yet both types of coordination being better than selecting unilateral cooperation.

The specification of players' primary and secondary goals constitutes a clear foundation and identification of players' preferences (Brams 1985, 120). Naturally, players' preferences derive from what they desire to achieve in the interaction. It is then sufficient to identify Row's objectives to obtain her preferences; Column's objectives remain identical to Row's. The Stag Hunt game matrix informs that Column's cooperation generates the best and the next-best outcomes for Row: CC generates a rank of 4 and DC where Row defects and Column cooperates generates a rank of 3.[6] Therefore, Row's primary goal is Column's cooperation. Therefore, the outcomes of CC and DC are more valuable for Row than those outcomes where Row cooperates and Column defects (CD) and Row defects and Column defects (DD). The primary goal of Row implies that {CC, DC} > {CD, DD}. Row has a secondary goal: reciprocity, that is, respond to cooperation by cooperation and respond to defection by defection. Consequently, the outcome CC is more preferable to DC, and the outcome DD is more preferable than CD. Row's preference ordering over the outcomes becomes: CC > DC > DD > CD. Column's preference ordering is the same as Row's. Column's primary goal is Row's cooperation so that {CC, CD} > {DC, DD}. Column's secondary goal is reciprocity. Therefore, Column's preference ordering becomes CC > CD > DD > DC.

The game represents coordination as a bifurcated process by its two equilibria with each representing an alternative coordination outcome. Mutual cooperation

---

[6] The first letter of the outcome refers to Rpw's choice and the second letter refers to Column's choice.

constitutes a Nash equilibrium and results for both in their best possible payoffs in the game: 4. Mutual defection also constitutes a Nash equilibrium, yet players obtain their next-worst payoffs of 2. Unilateral cooperation and defection do not lead to equilibria as each has an incentive to deviate from them given the other's choice. If Column defects, Row obtains the worst payoff of 1 in the game if she cooperates. Similarly, if Row defects, Column's unilateral cooperation generates the worst payoff of 1 for Column. The best reply of a player depends on what strategy she believes the other will select (Skyrms 2001, 32). Thus, if Row believes that Column will cooperate, then Row cooperates. If Column believes that Row defects, then Column would defect as well.

There exist five variants of Stag Hunt presented as "no-conflict games" by Rapoport and Guyer (1966).[7] Mutual cooperation yields the best payoff for both players in all five variants. The alternative equilibrium of mutual disagreement generates either the next-best or the next-worst payoffs of 3 and 2, respectively; it never constitutes the worst for both players. The five variants, namely, games 58, 59, 60, 62, and 63 are depicted below (Tables 4.8, 4.9, 4.10, 4.11, and 4.12):

Row's primary goal is to achieve reciprocation on the basis of mutual cooperation in the variant 58. Thus, CC and DD outcomes are more valuable than Row's unilateral cooperation outcome CD and Row's unilateral defection outcome of DC. Hence, for Row, CC > DD > {CD, DC}. Row's secondary goal is to be cooperative when there is no reciprocation. Therefore, CD is more valuable than DC for Row. Hence, Row's preference ordering is CC > DD > CD > DC. As to the Column player, she prefers Row's cooperation the most. Hence, for Column, mutual cooperation and Row's unilateral cooperation are more valuable than the outcomes reached by Row's defection: {CC, CD} > {DC, DD}. Column's secondary goal of reciprocation ranks the mutual cooperation outcome CC as more valuable than Row's unilateral cooperation outcome of CD. It also ranks the DD outcome as more

Table 4.8 Variant 58

Row		Column	
		C	D
	C	4, 4	2, 3
	D	1, 1	3, 2

Table 4.9 Variant 59

Row		Column	
		C	D
	C	4, 4	2, 2
	D	1, 1	3, 3

---

[7] These are games 58, 59, 60, 62, and 63 in Rapoport and Guyer. The game 61 corresponds to Stag Hunt as O'Neill (1987, 635) confirms.

Table 4.10 Variant 60

Row		Column	
		C	D
	C	4, 4	2, 1
	D	1, 2	3, 3

Table 4.11 Variant 62

Row		Column	
		C	D
	C	4, 4	1, 2
	D	3, 1	2, 3

Table 4.12 Variant 63

Row		Column	
		C	D
	C	4, 4	1, 2
	D	2, 1	3, 3

valuable than DC. Therefore, Column's preference ordering is CC > CD > DD > DC in the game.

In the variant 59 Row's primary goal is to achieve reciprocation under mutual cooperation. Therefore, CC > DD > {CD, DC}. Row's secondary goal is that if reciprocation fails, it is better to reply to defection by cooperation, that is, unilateral cooperation is better than unilateral defection. Hence, Row's preference ordering is CC > DD > CD > DC. Similar to Row, Column prefers to achieve reciprocation under mutual cooperation. Therefore, CC > DD > {CD, DC}. Column's secondary objective is the inverse of Row's secondary objective: unilateral defection is better than unilateral cooperation. Column's preference ordering is therefore CC > DD > CD > DC (the outcome of CD is Row's unilateral cooperation and therefore Column's unilateral defection, and the outcome of DC is Row's unilateral defection and Column' unilateral cooperation).

Row's primary goal in the variant 60 is to achieve reciprocation but under mutual cooperation. Therefore, for Row CC > DD > {CD, DC}. Row's secondary goal is to cooperate when there is no reciprocation, implying that CD > DC. Row's preference ordering is CC > DD > CD > DC. Column's primary goal is identical to Row's: establish reciprocity under mutual cooperation: CC > DD > [DC, CD]. Column's secondary goal is exactly the same as Row's: cooperate when there is no reciprocation; unilateral cooperation is better than unilateral defection, implying that DC > CD. Column's preference ordering under these goals becomes CC > DD > DC > CD.

Row's primary goal is Column's cooperation in the variant 62. Thus, {CC, DC} > {CD, DD} for Row. Her secondary goal is reciprocity. Hence

CC > DC > DD > CD as in the Stag Hunt game. Column's preference ordering is the same as in the variant 59. Thus, Column's primary goal is to achieve reciprocity in cooperation rather than defection so that CC and DD are better outcomes than unilateral cooperation and defection. Thus, for Column, {CC, DD} > {CD, DC}. Column's secondary goal is the choice of defection against Row's cooperation as being better than the choice of cooperation against Row's defection: CC > DD > CD > DC.

In the variant 63, Row's primary goal is reciprocation while reciprocation of cooperation is more valuable than the reciprocation of defection. If no reciprocation, Row's secondary goal is to achieve unilateral defection as more valuable than unilateral cooperation. Therefore, Row's preference ordering is CC > DD > DC > CD. As to Column, her primary objective is the same as Row's primary objective, but her secondary objective is the inverse of that of Row: CD > DC. As a result, Column's preference ordering is CC > DD > CD > DC.

Variations in players' primary and secondary goals do not affect equilibrium profiles. The mutual cooperation and mutual defection equilibria are equally likely under players' running the risk of ending up in mutual defection in Stag Hunt and all its variants. Psychological pressures operating on players' beliefs about reciprocation of cooperation and defection make mutual cooperation risky.

In general, some "primary reasons" shape players' preferences in interactions (Davidson 1963, 685). Primary reasons constitute the appealing aspects of reaching a mutual understanding to achieve an outcome of coordination. Players in VSGs who are IR scholars are assumed to evaluate a common semiotic approach is needed because it constitutes the appealing aspect of interacting over the meaning of abstract art. The game of Battle of Sexes illustrates that, unlike the game of Stag Hunt, failures of coordination generate no satisfaction for the players at all. This implies that, in for example, the Battle of Sexes game, IR scholars evaluating a meaning of a painting by Pollock and Rothko must rationalize more strictly their strategy choices. They would reason about their coordination as a process of agreeing on an appealing aspect of talking the same language over theoretical works of Wendt and Waltz through paintings by Pollock and Rothko, respectively.

Suppose now that the Stag Hunt game is presented as a perfect information game in extensive form (Fig. 4.3).

Fig. 4.3 Perfect information Stag Hunt game in extensive form

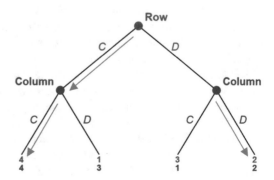

Suppose that Row starts the interaction by choosing between C and D. Column's and Row's reactions are in bold. Column will react to C by C and D by D. Row foreseeing Column's reaction and reasoning back will select C to avoid the DD equilibrium. Therefore, perfect information of a player who acts after being informed of the other player's choice will serve to eliminate the DD equilibrium. Thus, perfect information of the player who moves second works for the interest of both players as the outcome will be CC where both players obtain their best. The equilibrium will be mutual cooperation regardless of who moves the first in the game. The Nash equilibrium of DD is eliminated by the condition of perfect information so that the player who moves first believes that the player who moves last will commit no mistake of responding to C by D. Hence, the elimination of the disadvantageous equilibrium of DD becomes possible by the first moving player's counting on the rationality of the last moving player. Stag Hunt game in perfect information in extensive form has a unique equilibrium in mutual cooperation in contrast to the game in strategic form reflecting the assumption of imperfect information so that no player is informed of the other's choice.

4.3.3   Battle of Sexes

The game of "the battle of sexes" is a frequently used example of coordination problem among humans (Luce and Raiffa 1957, 90–94). Players are under the pressure of avoiding unilateral cooperation and defection like in the game of Stag Hunt and all its variants. The Battle of Sexes game differs from the Stag Hunt game and all its variants as players obtain zero satisfaction in coordination failures. Players must coordinate their strategy selections constituting Nash equilibria that would benefit one player more than the other and vice versa unlike in the game of Stag Hunt.

To illustrate the coordination problem the game of Battle of Sexes poses, suppose that a couple in love want to spend a night's entertainment together. They evaluate two options: see a sci-fi movie or go to a rock concert together, yet one of them prefers the sci-fi movie and the other the rock concert. Players' (partners') preferences stem from living together generating a social environment of the couple like the couple's past interactions with each other and inner personal sensations and pleasures. Sensations of watching a sci-fi movie and listening to a rock group in a concert hall vary in intensity from one player to the other. Nevertheless, the couple in love is under the constraint to be together. They avoid being separated. Togetherness is the source of the couple's strategic interdependence. If players were alone, each would manage to spend his/her leisure time without having to think about what the other prefers; however, they are not alone. Any disagreement emanating of partners' different choices means separation and resulting frustration suffered from separation. A meeting of minds is necessary to avoid separation.

Luce and Raiffa (1957, 91) offer another way to eliminate an equilibrium in battle of sexes where a man and a woman interact who decide to go either to a prize fight or to a ballet and man prefers prize fight while woman prefers ballet:

Table 4.13 Generic Battle of Sexes game

Row		Column	
		Movie	Concert
	Movie	a, b	0, 0
	Concert	0, 0	c, d

> If, in a preplay discussion, the man says he is already committed to the prize fight and demonstrates his intention of going by producing the ticket he has already purchased, this may cause the woman to submit to his will [...] But to some spirited females, such an offhand dictatorial procedure is resented with sufficient ferocity to alter drastically the utilities involved in the payoff matrix. Preplay communication is considered outside the game structure of the payoff matrices, yet in some cases it may result in a radical change of one player's preference pattern and therefore of the payoff matrix.

They add that "we shall suppose the payoff matrix remains invariant during the negotiations" (ibid.). Thus, preplay discussion offers possibilities of alternative moves reducing the number of two equilibria to a unique equilibrium.

The generic Battle of Sexes game is given below (Table 4.13):

The generic Battle of Sexes game generates variants like $a > b$ and $c < d$ or $a < b$ and $c > d$. The Battle of Sexes game transforms into a Pure Coordination game if players do not care whether they go to the movie or concert together so that they evaluate being together at a movie or a concert equally, then $a = b$ and $c = d$. If both players prefer to be together but one enjoys being at the movie the most unlike the other who prefers concert the most (or vice versa) then $a > b, d > c$ as discussed by Luce Raiffa (1957, 90) (Table 4.14):

It is also possible that, for example, both partners enjoy being together at the movie more than being in the concert, so that $a = b > c = d$ or they enjoy being together at the concert, more than being together at the movie, so that $a = d < c = d$. These last two cases imply that being together at the movie and at the concert become focal, respectively (Schelling 1983). It is also possible that going together to a concert or movie creates a considerable amount of satisfaction for one of the players, but this does not change the equilibria.

Each version implies two equilibria: either both players go to the movie or to concert together yet players' utilities vary. Game theory does not answer the question of which of these two equilibria will be realized. Keohane (2000, 128) states that "coordination takes place as a result in part of the ideas people have not only about their own beliefs but about the beliefs of others." The game illustrates that Row and Column's beliefs about mutual beliefs do not determine whether the couple will go to the movie or concert together. Some extra-game theory assumptions such as commitment of a player to a strategy and her communicating her choice to the other remain a possibility of the elimination of one equilibrium in favor of the player who moves first by committing herself to a choice constraining the other's choices (Schelling 1983, 47). For example, Row can inform Column that he is committed to the choice of movie letting Column to choose between movie so that she will get a utility of *1* or *0* if she chooses concert. Column then will select movie as well to be together with her partner. The game illustrates that players expect to

Table 4.14 Battle of Sexes

Row		Column	
		Movie	Concert
	Movie	2, 1	0, 0
	Concert	0, 0	1, 2

Fig. 4.4 Perfect-
information battle of sexes
game in extensive form

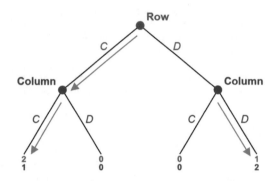

reciprocate choices regardless of the level of satisfaction they get from the equilibria. Players' worst outcomes represent failures of coordination. They interact under strong incentives of reaching a meeting of minds, that is, concordant expectations about mutual strategy choices.

Preplay communication is a powerful remedy to reduce the impact of strategic uncertainty upon players' minds. Players can discuss, for example, where they will go together for a night's entertainment in games of battle of sexes before they decide, or in dress-color game they can discuss what choices they will make before they select colors. Therefore, communication is helpful for players to harmonize their mutual beliefs about how they will act. Perhaps battle-of-sexes players have played the game frequently in the past and achieved a balanced relationship by accepting one's proposition first and accepting the other's proposition in the second play and so on. They might have agreed that a chance move such as drawing a red or a black card from a deck of cards to decide where to go that night. Or, in the dress-color game, Barbara and Alice accept that Alice wears blue and Barbara green and so on in successive social gatherings. There can be many other opportunities preplay discussion might create for players to coordinate. For example, imagine that players meet for the first time and discuss meanings they concoct for propositions Waltz and Wendt offer in the Discipline in a VSG. They might have a history of interactions, or they could find an interest in following recommendations a neutral referee formulates for them amounting to the concept of correlated equilibrium (Aumann 1987).

The Battle of Sexes game in extensive form given below illustrates how one of the two equilibria will be eliminated (Fig. 4.4):

Row looks forward and realize that regardless of its strategy selections, Column reciprocates her choice. Column reacts against Row's choice of C by C and Row's choice of D by D. As the outcome CC yields a payoff of 2 and the outcome DD a payoff of 1, Row will choose C to obtain a utility of 2. As result, one equilibrium survives in the interaction: CC; DD equilibrium will not be played. All coordination games covering pure coordination, battle of sexes, and stag hunt in extensive form and perfect information generate a unique Nash equilibrium found by backward induction. Now, we turn to coordination problem as modeled by Saussurean games.

Chapter 5
Saussurean Games

Saussure (1996, 98) asserts that a linguistic sign emanates from an association between a concept and an acoustic image, the two elements of a sign:

> Le signe linguistique unit non une chose et un nom mais un concept et une image acoustique. Cette dernière n'est pas le son materiel, chose purement physique, mais l'empreinte psychique de ce son, la representation que nous en donne le témoignage de nos sens; elle est sensorielle, et s'il nous arrive de l'appeler "matérielle", c'est seulement dans ce sens et par l'opposition à l'autre terme de l'association, le concept, généralement plus abstrait.

Saussure's definition of linguistic signs finds a niche in visual terms. Following Saussure, I argue that a visual sign is not a painting but an IR concept the painting represents. The painting helps an IR scholar to form an association between itself and an IR proposition in her mind. Whether the painting is qualified as abstract or not matters as long as the sign depends on human senses of a correspondence in purely abstract terms. The painting becomes a visual signifier provided that it helps the formation of a mental correspondence between itself and propositions by Waltz and Wendt. The propositions find home in the realm of abstract art as a result. Not all paintings are helpful; helpful paintings aid IR scholars to nail down theoretical essentials. Scholars' task is not an easy one because contours (if they ever exist) of artistic abstract thinking subsume linguistic theoretic abstractions.

Abstract paintings containing fewer essentials ease the task of forming visual signs for IR propositions. Pollock's drip-painting works can be assessed as quite complex but if one evaluates them in terms of curls of colors, then it is possible that one cognizes the simplicity of the work. The rhythm of continual curls makes the redundancy of color shifts relevant even if shifts subsume different colors. Color curls can be interpreted as ever-turning paths in life. It is then up to scholars to make a correspondence between theory and painting along the abstract-abstract axis.

IR scholars are players who interact about the sign quality of a painting in a Saussurean game. Each decides whether a painting helps or not as a signifier for a theoretical proposition that is the signified. The sign is the association of the painting with an IR concept or proposition. Following Saussure, players' preferences in

© Springer Nature Switzerland AG 2023
S. Ş. Güner, *Art and IR Theory*, Mathematics in Mind,
https://doi.org/10.1007/978-3-031-32342-3_5

VSGs are nothing but subjective evaluations of whether the painting is "a material constituent (written trace, phonic element)" for the "immaterial idea or concept" (Krauss 1981, 15). For example, a player can select a Rothko color-field painting as a signifier of Waltz's figure because she believes that the artwork constitutes a visual representation of Waltz's figure. The choice of signifier means that there exists a subjective exactness of painting-proposition connection for a player. It corresponds to the player's idea that the figure visually represents Waltz's representation of his mental picture of systems approach in IR. The function of the signifier is to reveal structural constraints through color interactions (Albers 2013) instead of the two arrows in the figure.

Croce (1965, 22) asserts that "all this world is intuition." IR scholars rely on their intuitive capacities when they interact. Games imply that each player ponders about mutual intuitions. Players communicate to understand each other and exchange intuitions about the correspondence between a painting and an abstract IR proposition. The communication is at intuitive, mental, and individual levels helping players to coordinate choices. The communication is not easy as paintings are not helpful compared to an advertisement for a car brand (Fiske 2011, 13). Therefore, players must think hard about how paintings and theoretical propositions mutually connect and exchange their intuitions about these connections.

Suppose that players who are IR scholars have formed specific signs in their minds. Players then come together and exchange their beliefs about artistic aspects of either Rothko or Pollock which they perceive as if they reflect or correspond to propositions by Waltz and Wendt on structural constraints and nonunique dynamic anarchies, respectively. Preplay discussion corresponds to this exchange between players before they select strategies. It is a debate between players over signs and ideas they form and a process of interactive meaning generation through which players may revise and update their sign convictions. The debate clarifies for players what outcomes and therefore what strategy profiles should be avoided. Players' failures in coordination constitute a central problem to solve in all coordination games similar to equilibria indicating an interactive strategic stability in these games.

The Pure Coordination, the Stag Hunt, and the Battle of Sexes games given in the previous chapter constitute Saussurean games with Row and Column possessing two strategies in strategic form: "accept" (A) and "reject" (R). The strategies correspond to players' acceptance and rejection of the painting as a signifier for the theoretical proposition that is the signified.

5.1 Nash Equilibrium

Equilibrium conditions in the generic Saussurean game are the following:

1. *(A, A)* pair of strategies constitutes a Nash equilibrium if and only if $a \geq 0$ and if $b \geq 0$;
2. *(A, R)* pair of strategies constitutes a Nash equilibrium if and only if $0 \geq c$ and if $0 \geq b$;

3. *(R, A)* pair of strategies constitutes a Nash equilibrium if and only if $0 \geq a$ and if $0 \geq d$;
4. *(R, R)* pair of strategies constitutes a Nash equilibrium if and only if $c \geq 0$ and if $d \geq 0$.

It is assumed that a, b, c, and $d \geq 0$. Therefore, the generic Saussurean game leads to two Nash equilibria either both prefer coordination on A or on R. Both equilibria indicate a development of a common semiotic language in Saussurean terms. The Nash equilibria remain constant while utilities change: *{(A, A), (R, R)}*. Thus, while players' payoffs change, players either accept or reject that Rothko's and Pollock's paintings signify structural constraints and dynamic anarchies, respectively. As a result, utility changes in *(A, A)* and *(R, R)* do not prevent the emergence of stable mental worlds as long as equilibrium conditions are fulfilled.

Both equilibria mentally transpose paintings' artistic meanings into the domain of IR theories. The equilibrium of *(R, R)* means discoordination implying players' convergent interest in rejecting the paintings as signifiers of structural constraints so that players oppose a visual semiotic language in the Discipline. It signals that there is no hope in developing a common sign concerning structural constraints and dynamic anarchies. Players' mental calculations are illustrative of Nash equilibrium strategies. For example, the Row can be viewed as reasoning as: "I will select A because I expect the Column to play A as she expects me to play A." Hence, as the Row expects the Column to expect the Row to expect the Column to play A, the equilibrium of *(A, A)* obtains. The same reasoning equally applies to all equilibria of the three games. Consequently, the dual existence of equilibrium demonstrates that players can fail to cooperate even if both prefer to reach an agreement by showing how players' strategic uncertainty about mutual choices, that is, how "each player's best choice of action depends on the action he expects the other to take, which he knows depends, in turn, on the other's expectations of his own" (Schelling 1980, 86) can prevent the development of a common semiotic language. Thus, players' interest in achieving an agreement does not suffice to attain stability in development of a sign under strategic uncertainty.

No sign exists outside players' minds. The mental calculations based on the game rules and players' subjective preferences back the antirealist quality of semiotics. The sign, while stable in game theoretic sense, cannot be made false or true by the facts in the world. In the generic Saussurean game, if a, $b > c$, d, it is obviously better for players to agree on the sign with both obtaining higher payoffs compared to the joint disagreement. Yet if one believes that the other is not so welcoming the idea of a specific painting as a signifier and therefore it is better to select R, one will realize that going for a joint agreement invites only both receiving zero payoffs provided her belief about the other is incorrect. It is instructive that both equilibria stem out of players' mutually consistent expectations of mutual strategy choices about the immaterial idea of Rothko's and Pollock's paintings signify structural constraints and dynamic anarchies, respectively. It is possible for the players to delineate the *(A, A)* equilibrium as the focal point of the game since players' expectations could converge on it as it is twice better that the *(R, R)* equilibrium.

Nevertheless, players' suspicions about each other's artistic perspectives and ability to construct a visual sign about a theoretical sentence can lead to the collapse of the *(A, A)* equilibrium. Saussurean games come in different varieties, now we turn to their analysis.

5.2 Saussurean Stag Hunt Game

The game below is a Saussurean Stag Hunt Game (Fig. 5.1):

The game is a copy of the original Stag Hunt game with the only difference of strategies referring to the semiosis process. Players make their strategy choices in imperfect information condition in the game. Hence, a player chooses accept and reject without knowing the choice of the other player; they move simultaneously.

Saussurean games posit that players are IR scholars who do not reject at the outset the usefulness of abstract art in the Discipline. They model coordination problems where players' strategy sets consist of accept and reject decisions that a painting constitutes a signifier for a proposition of Waltz or Wendt, that is, signified elements of signs. Saussurean games can be expressed in strategic and extensive forms. In strategic form, they are 2 × 2 games so that two players interact each possessing two strategies. In extensive form, they posit a player who moves first by choosing between actions of accept and reject. The player who moves next chooses between actions of accept and reject. She can be informed or not of the previous action the first moving player has taken before. The game ends once both players made their selections.

If players fail to coordinate, they obtain lesser payoffs compared to coordination outcomes. Two coordination outcomes exist: either both players accept or reject the signifier quality of a painting for a theoretical proposition. Common acceptance of a painting as being a signifier for a theoretical proposition reflects the emergence of a common semiotic language between players. Common rejection means that players would rather prefer an alternative painting to be considered as a signifier; it does not mean a definite rejection of a theoretical communication where abstract art constitutes an interpretive ground. If both players agree that a painting in question cannot constitute a signifier, it means that both prefer to see a different painting to discuss the sign quality of the artwork for structural constraints and nonunique dynamic anarchies.

The two equilibria Saussurean games in strategic form imply that either players reach an agreement on the sign or not. If players coordinate and reach an agreement,

		Column	
		A	R
Row	A	4, 4	1, 3
	R	3, 1	2, 2

Fig. 5.1 Saussurean Stag Hunt Game in strategic form at ordinal level

then the equilibrium implies a fruitful communication. Players who disagree on the meaning of a painting do not necessarily oppose a visual semiotic language leading to a common visual semiotic conceptualization of IR theory sentences. The outcome emanating of both players' rejection is instructive in the sense that players can oppose a painting because they evaluate it as not being meaningful in the direction of development of a common language. They might find an alternative painting as a more useful artwork for that purpose. The (R, R) equilibrium can be interpreted as players' rejection of the idea of paintings constituting any signs. Perhaps players refuse to talk about what they understand of Waltz's and Wendt's theoretical propositions through paintings; such a semiotic idea is not attractive for them.

It is possible to interpret VSG interactions in terms of metaphors. Players facing strategy choices reflect players' views about whether the painting constitutes a vehicle to appreciate Waltz's proposition, the tenor. The tenor-vehicle relation connotes a metaphor where "the tenor is the thing treated and the vehicle is the treatment" (Kennedy 1982, 589). Therefore, in terms of Saussurean semiotics, strategy choices can imply visual metaphors of a Rothko's painting represent or not Waltz's proposition. The signifier becomes a painting by Pollock or Rothko, and the signified becomes theoretical concepts of dynamic anarchies and structural constraints, respectively (Güner 2019, 2021). A unilateral rejection or acceptance is less valuable or costlier than a common understanding in refuting a painting jointly.

Saussurean signs suffer two principal problems in the context of VSGs. First, there is the difficulty of forming visual signs on the basis of verbal ones (Hatt and Klonk 2006). The priority is given to verbal signs in Saussure who founded semiology on the basis of language. Visual signs are less arbitrary compared to verbal signs. To illustrate, cat names, say "Joy" or "Dido," are not necessarily connected to cats in general. It is impossible for someone to draw a cat picture (or mentally form a picture of a cat) hearing the word "Joy" and "Dido" (except humans the cats Dido and Joy own). However, if a photograph of Joy or Dido to anyone else is shown, they will be recognized as cats the photographs depict. Joy and Dido photographs then become signifiers (and icons) of Joy and Dido. Thus, players' approval or disapproval of signs reveal conditions of players' emphatic and intersubjective interpretations of the paintings facilitating their development of signs out of Waltz's and Wendt's statements. If players have a full knowledge of Constructivism or Structural Realism and contemplate the same painting and agree on the meaning of what is represented and how it is represented, we can affirm that a visual semiotic language in IR is possible. The language has two consequences for the Discipline. First, players' interactions reveal what they understand of Waltz's and Wendt's IR approaches and therefore what features of these two approaches they disagree about. Second, interactions clarify uninvestigated implications and new aspects and conclusions about Structural Realism and Constructivism.

Saussurean games can be analyzed as one-shot games, finitely or infinitely repeated games. The equilibrium analysis is prone to offer strategy choices that are best replies against each other in each case. If, for example, a Saussurean game in the form of game of Battle of Sexes played once, it implies that unilateral acceptance or rejection of are impossible. It is impossible that one player accepts and the other rejects the signifier quality of a painting. If it is repeated finitely many times, then the play of a Nash equilibrium at each round becomes the solution: either both accepts or rejects at each round or one round of acceptance followed by a round of rejection or vice versa. Each such sequence constitutes strategic stability (Watson 2008, 257–262).

If both players' choose to agree that the painting in question constitutes a signifier for a proposition, then players obtain the highest possible payoff in the game. No player has an incentive to shift from agree to do not agree against the other player's selection of agree. The strategy profiles of "agree, disagree" and "disagree and agree" do not constitute Nash equilibria because the strategies of agree and disagree are not best replies against each other: against agree disagree strategy yields a payoff of 3 which is smaller the highest utility of 4 in the game deriving from the reciprocation of the agree strategy for both players. If both players accept or reject, then they reach coordination. They obtain positive utilities of a and b provided that both players agree on the sign and c and d provided that both players disagree. The worst outcomes of unilateral acceptance and rejection generate utilities of 1 for both players.

The highest satisfaction and reward by agreeing about the sign quality of a painting, yet they might also end up in a Nash equilibrium by rejecting the sign quality of the painting. The rejection may imply players' rejection of the semiotic idea that paintings can imply meanings for either Waltz's or Wendt's proposition. The game of Stag Hunt reminds that even if both players prefer cooperation by agreeing together that the painting in question stands for the proposition either by Waltz or Wendt the most, they can fail to do so and can find themselves in the alternative equilibrium of mutual disagreement. The variants 59, 60, and 63 attribute the next-best payoffs for both players in disagreeing the painting not constituting a signifier. Thus, players who are after developing a semiotic approach toward the propositions by Waltz and Wendt prefer to see an alternative painting as a signifier. The alternatives 58 and 62 imply that players get different satisfactions from disagreement. The appealing aspect of the mutual agreement rewards players equally in all Stag Hunt variants. The outcome of mutual disagreement generates either equal payoffs corresponding to next-best outcomes or unequal ones where one gets the next-worst payoff or 2 and the other the next-best payoff of 3 and vice versa.

The game of Stag Hunt represents the stability of players' coordination and rejection of the painting as constituting a signifier for either Waltz's or Wendt's theoretical proposition. It represents coordination in two opposite directions. Players' agreement means a positive coordination. Players' disagreement is a negative coordination; it means that players agree that an alternative painting would better serve as the sign for the two propositions; it does not constitute a failure of coordination. The "disagree, disagree" equilibrium represents a lower satisfaction for both players compared to the alternative one yet it is better than a coordination failure.

VSGs generate explanations on the basis of their equilibrium or equilibria indicating stable interpretations, therefore no player would unilaterally deviate from an equilibrium. The equilibria incorporate intertwined subjectivities and perceptions producing stability. They are helpful in detecting what theoretical features works of Waltz and Wendt hide, keep, or withhold. For example, Structural Realism is criticized to be a static theory. Yet we reach a result supporting that Structural Realism is not static but is able to feature dynamic traits at system level. A luminosity perceived as stemming from an interaction between two areas of different colors of Rothko painting constitutes a sign of a dynamics Structural Realism captures while it is castigated for being a static IR theory. Structural Realism becomes liberated from its static interpretive cage in which it is imprisoned. As to Constructivism that propose three types of security systems, some equilibria of VSGs demonstrate that there is an infinite number of anarchies according to Constructivism.

If players reach an agreement, it should be self-enforcing; no one would deviate from it while others stick to it. Thus, the agreement should be a Nash equilibrium (Kreps 1990, 411). Saussurean (and Peircian) games are solved for Nash equilibrium that corresponds to a pair of strategies such that neither player has any incentive to deviate unilaterally from her strategy provided that the other sticks to hers because she would do worse or not better if she deviates.

Stag Hunt game in extensive form eliminates the (R, R) equilibrium by its backward induction solution. The solution of the selection problem is real and exists. The best way to eliminate one of the two equilibria in coordination games is to allow players to move sequentially. Thus, in short, the real test of a preplay agreement comes through a Saussurean game in extensive form. Column player moves after Row being informed of Row's choice; she has perfect information. It is possible to demonstrate a first-mover advantage by a Saussurean perfect-information game in extensive form:

The equilibrium of (A, A) emerges as a result of players' sequential choices and their ability to look forward and reason back. Players rely on the rationality of the others: the player who moves last choose A against A and R against R because these are optimal reactions to the player who starts the interaction, the first-moving player foreseeing last moving player's reactions make up her mind and selects the action A leading to the best utilities for both players in the Nash equilibrium of (A, A). Therefore, it is in the common interest for the players to interact sequentially to avoid the (R, R) equilibrium that is suboptimal for both of them.

A close inspection of the game shows that the last moving player, say Column, at both of her information nodes has dominant actions so that her strategy is "if the first-moving player chooses A respond by A and if she chooses D respond by D." Being completely informed of the game rules and therefore that the last-moving player has dominant actions, the starter player's, say Row's, action of A dominates her action of D. Column is prepared for any choice Row might take at start. There is nothing wrong for Row to count on Column's ability to choose her dominant action.

5.2.1 Saussurean Battle of Sexes Game

The Battle of Sexes game in Saussurean sense given in Fig. 5.2 posits two strategies for each player: "accept" (A) and "reject" (R) as given below where $a > b > 0$.

Suppose that $a = 2$ and $b = 1$ like in the original game offered by Luce and Raiffa (1957, 90). The game implies that players have a common interest in avoiding (A, R) and (R, A) outcomes generating negative payoffs for each. The failure of coordination is costly for both. Row obtains a utility of 2 and Column a utility of 1 in the equilibrium of (A, A) and that they obtain 1 and 2 in the equilibrium of (R, R). Row prefers the (A, A) equilibrium and Column prefers (R, R) equilibrium. It is better to reach an agreement or equally a disagreement about the possibility of a visual semiotic language. The equilibria reflect players' opposed preferences with respect to the development of forming signs on the basis of paintings. The problem however remains: who will get the payoff of 2 and the payoff of 1? If both accept that the painting constitutes a signifier for a proposition, the Row becomes more satisfied; otherwise if both reject, Column becomes more satisfied. Players' objective is to reach coordination but on their own preferred terms. This is the element of conflict the game introduces between the players (Fig. 5.3).

5.2.2 Mixed-Strategy Nash Equilibrium

The mixed-strategy Nash equilibrium permits players to accept and reject a painting as a signifier probabilistically. Players do not select strategies in a mixed strategy equilibrium by certainty. This relieves them of the pressure of going either one or

Fig. 5.2 Saussurean Stag Hunt Game in extensive form

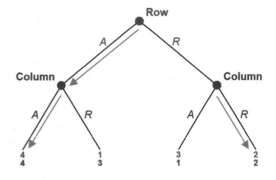

		Column	
		A	R
Row	A	a, b	0, 0
	R	0, 0	b, a

Fig. 5.3 General Saussurean Game

		Column		
		(q) A	(1 – q) R	The Row's payoff
Row	(p) A	2, 1	0, 0	2 q + 0 (1 – q)
	(1 – p) R	0, 0	1, 2	0 q + 1 (1 – q)
	The Column' payoff	1 p + 0 (1 – p)	0 p + 2 (1 – p)	

Fig. 5.4 Mixed Strategies

the other way because there might be a solution which is jointly pleasing under concordant interests of deciding whether the painting is a signifier or not. Thus, players avoid prospects of discoordination due to unilateral acceptance or rejection of paintings as signifiers. They would not be willing to risk coordination failures. Dixit and Nalebuff (1991, 189) succinctly state why players would use mixed strategies in their interactions: "the reason for mixing one's moves arises from a failure of coordination."

The mixed equilibrium is unique. It can be interpreted as a solution to the problem of alternative coordination outcomes. A mixed-strategy Nash equilibrium emanates from players' interdependent indifferences. All games have mixed-strategy equilibria in which players reject and also accept that a painting constitutes a signifier by some likelihoods that would be called as "fuzzy signs."

In the Saussurean Battle of Sexes game Row plays A with p and R with $1 - p$ probabilities, and, similarly, the Column plays A with q and R with $1 - q$ probabilities as given below (Fig. 5.4):

Row' payoff to playing A is $2 q + 0 (1 - q) = 2 q$ and R generates him a payoff of $0 q + 1 (1 - q) = 1 - q$. Column selects such probabilities of q and $1 - q$ that make Row indifferent between playing A or R by certainty. Thus, Column's mix should satisfy the condition of equality of Row's payoffs to A and R. The condition yields the equality of $2 q = 1 - q$ and therefore $q = 1/3$. Similarly, the Column' payoff to playing A is $1 p + 0 (1 - p) = p$ and R generates her a payoff of $0 p + 2 (1 - p) = 2 (1 - p)$. Similarly, the Row selects such probabilities of p and $1 - p$ that make the Column indifferent between playing A or R by certainty. Thus, Row's strategy mix should satisfy the condition of equality of Column's payoffs to A and R. The condition yields the equality of $p = 2(1 - p)$ and therefore $p = 2/3$. Consequently, the mixed Nash equilibrium becomes $p = 2/3, q = 1/3$. The result informs that if Row accepts that the painting constitutes a signifier for an IR proposition with a probability of $2/3$ and rejects with a probability of $1/3$, and, Column accepts that the painting constitutes a signifier for an IR proposition with a probability of $1/3$ and rejects with a probability of $2/3$. Row and Column get, as a result of their mixed choices equal payoffs. Row obtains $(2/3)(1/3)(2) + (2/3)(2/3)(0) + (1/3)$ $(1/3)(0) + (1/3)(2/3)(1) = 2/3$; Column similarly obtains: $(1/3)(2/3)(1) + (2/3)(2/3)$ $(0) + (1/3)(1/3)(0) + (1/3)(2/3)(2) = 2/3$. Therefore, the equilibrium benefits Row and Column equally. Mixed strategy equilibrium is useful in the sense of equal benefits to players who randomize their choices when there is problem of coordination.

What patterns the mixed-strategy equilibrium reveals for this game or what do we learn from it? What does the condition $p = 2/3$ and $q = 1/3$, that is, Row and Column respectively select "accept that the painting constitutes a signifier" two-third and one-third times when they interact teach us? Rubinstein (1991, 913–915) offers three alternative interpretations. First interpretation is the view that the Saussurean game is played among IR scholars which become divided into those who accept and reject the sign over time. Ultimately, two-thirds of the population accept, and one-third of the population reject that the painting is constitutive of sign for the theoretical proposition in question. Thus, coordination takes two alternative directions along with precise population proportions with a proportion of scholars, two-thirds of them become acceptors and one-third of them become rejectors. The population of IR scholars involved in the coordination game above becomes divided, in short.

The second interpretation aligns with a more individualistic view: there are factors belonging to players' own private subjectivities and randomization between acceptance and rejection of signs (Rubinstein 1991, 914). Some IR scholars might be for or against a communication using paintings as signifiers of Waltz's and Wendt's propositions. IR scholars might not believe that paintings are helpful in their communication; they might not be hundred percent favorable to the making meanings of paintings for structural constraints and nonunique dynamic anarchies. A propositional attitude as a mental phenomenon formulated by McGinn (1982, 8) implies that both players do not believe or perceive that the painting constitutes a signifier. Beliefs and perceptions are individual traits which shape players' preferences. Therefore, beliefs and perceptions come out of IR scholars' co-existence. The argument gains support in the context of the Discipline divided in terms of methodological approaches and geographical terms (Tickner 2011). Two-thirds of IR scholars might agree that a visual semiotic language is possible in the Discipline and one-third of them either oppose the language or prefer an alternative painting to discuss theory. Hence, a change in the painting can also change the mixed-strategy equilibrium in the Saussurean coordination game.

The third interpretation shifts attention from players' own mental and private subjectivities to attention they pay to the other player in the game (Rubinstein 1991, 915).[1] The third interpretation views mixed strategies in a VSG as, for example, the belief of Row concerning Column's strategy choice and vice versa. Thus, a mixed-strategy equilibrium becomes a pair of common knowledge expectations; one element referring to those expectations Row holds about Column's acceptance and rejection and the other referring to Column's expectations about Row in a similar fashion. The interpretation has strong connections with and implications in the philosophy of mind we will turn to in the last chapter.

---

[1] Rubinstein informs that the third interpretation is due to Aumann: Robert J. Aumann. 1987. "Correlated Equilibrium as an Expression of Bayesian Rationality." *Econometrica* 55: 1–18.

5.2.3 Correlated Equilibrium

The concept of correlated equilibrium by Aumann (1987) offers an alternative solution to the coordination problem. Correlated equilibrium concept assumes a device, a third party that is trusted by both Row and Column that makes a signal observed by both players. The signal is, say, a flip of fair coin and the resulting outcome like heads and tails. Both sides of the fair coin have equal occurrence likelihoods of ½. Players use the experiment outcome to determine their strategy choices. For example, if the coin flip gives heads, Row and Column choose A; otherwise, if the coin flip gives tails, Row and Column choose R. Both players observe the outcome of the coin flip and select strategies. Row and Column ignore mutual choices; however, either Row or Column would not have any incentive to deviate, because each knows that the other sticks to the proposal the third party makes. If, for example, Column deviates from A to R after the outcome of flip is heads, then she obtains a zero payoff given that Row plays according to the signal and selects A. If Row deviates from R to A after the outcome of flip is tails, then she obtains a payoff of zero given that Column plays according to the signal and selects R accordingly. Therefore, correlated equilibrium is self-enforcing; no player would deviate from the toss-of-coin plan and do better unilaterally as in a Nash equilibrium. As to the coin toss, it functions as a mediation device in an uncertain environment where the question of how to coordinate choices about a painting serving as a signifier or not for Waltz and Wendt propositions is the main issue of the interaction.

What do players obtain out of the correlated equilibrium? Row obtains the following utility following the coin flip: $(½)(2) + (½)(1) = 3/2$. The utility of Row emanating of (A, A) is 2 occurring with the outcome of heads and the utility emanating of (R, R) is 1 occurring with the outcomes of tails. Similarly, Column obtains 1 with the probability of ½ if heads and a utility of 2 with the probability of ½ if tails comes out. Thus, Column's expected payoff is $(½)(1) + (½)(2) = 3/2$. The result is informative because the correlated equilibrium generates for each player a payoff of 3/2, a payoff which is strictly higher than the total payoff of 2/3 the mixed-strategy equilibrium generates. The elimination of discoordination prospects by the correlated equilibrium benefits both players equally in the game. No player gets exactly what she wants but at least no player suffers of prospects of coordination failures. The correlated equilibrium is "efficient" as the players' total payoff of $3/2 + 3/2 = 3$ is equal to the total payoff players obtain in successful coordination: $1 + 2 = 3$ (Gardner 2003, 88). Thus, the possibility of a trusted third party, a device, a mechanism, or an instrument that lies beyond the formal rules of the interaction, can improve players' satisfaction in Saussurean terms. Players go beyond linguistic meanings of IR theory propositions through their coordination over art, that is, a meaning-making process through Rothko and Pollock paintings.

5.2.4   Commitments

Schelling proposes commitments to solve the coordination problem. In any Saussurean coordination game "whoever can first commit himself wins" (Schelling 1980, 47). For example, if Row commits to *A* first in the game in Fig. 5.5 she forces Column to select *A* as well, and similarly if Column commits herself to *R*, then she forces Row to select *R* as well. The problem is to establish a clear communication channel so that players are informed of commitments. If players agree that a painting constitutes a signifier or not and both obtain equal utilities by selecting *A* or *R*, the commitment of one benefits the other.

A player's moving first corresponds to moving first in a coordination game. Players' sequential moves illustrate Row's moving first corresponds to her ability as the game in extensive form models:

The game is solved by backward induction. The solution implies that Column, as the last moving player, responds to Row's choice *A* by *A* and responds by *R* against *R*. Once Column's reactions to *A* by *R* and to *R* by *A* are eliminated, Row will choose *A* but not *R* to obtain her utility of 2 stemming from the outcome *(A, A)* instead of obtaining her utility of 1 stemming from the outcome of *(R, R)*. Column cannot commit herself to *R* because if Row chooses *A*, she has to respond by *A* to obtain a utility of 1 instead of 0. Column's commitment to *R* is not credible. Row does not have to commit to *A* because the structure of the game guarantees her the best utility as she moves first the game. The backward solution implies that the painting in question constitutes a signifier due to Row's advantage in her reaction to the last moving player's choice. It is possible that the utilities of *(A, A)* and *(R, R)* interchange so that the player who moves first obtains a utility of 2 from *(R, R)* and, as result, both players coordinate their choices on the rejection of the painting as a signifier. Players would then compete for the position of the first mover in the game.

Yet, players would also negotiate on artistic aspects of either a Pollock or a Rothko painting how would these artworks signify theoretical sentences before they decide who moves first in the game. They would zero in parts of the paintings helping them to construct correspondences in terms of meanings they create between art and IR theories. They "paint" IR theories by interacting with paintings; they become "fake" artists in turn. They express each other how they assess paintings which are

Fig. 5.5 Perfect-information battle of sexes game in extensive form

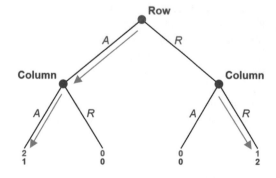

expressions of ideas of Rothko and Pollock as signifying Waltz's structural constraints and Wendt's dynamic nonunique anarchies, respectively. Players exchanging their feelings about art versus theory correspondences explore avenues, roads to leading to new meanings of Structural Realism and Constructivism. These exchanges are loaded with pure intuition and apprehensions. Hence, the words of anarchy and structural constraints are filled up with abstract art elements.

Neither Pollock's nor Rothko's art can be part of a meaningful communication unless the artists' paintings constitute a basis of interpretation of some theory. Players interpret the artworks and try to affect their mutual states of mind through exchanges of messages and correspondences they form between paintings and theories. An equilibrium emerges after those negotiations on artistic aspects and deciding who moves first given players' payoffs. Hence, if players can engage in preplay discussions, they become able to coordinate on one equilibrium out of the two at the expense of one of them. However, regardless of coordination on *(A, A)* or *(R, R)*, a common view and semiotic language emerges from the game. Each equilibrium tells a story constituting a myth (Barthes 1972).

5.2.5   Bayes-Nash Equilibrium: One-Sided Incomplete Information in Saussurean Games

The assumption of complete information sounds as quite distanced from empirical realities. Players cannot be aware of all "nuances in tastes and whims of other players" (Aumann 1985, 19). Consequently, it is worth to study implications of players' uncertainties about mutual incentives. Incomplete-information games study such complexities of limited information. They incorporate Bayesian thinking describing mental processes and issues to some extent (Davidson 1980).

For example, in the dress-color game, if Alice or Barbara or both misjudge preferences, then they become players of an incomplete information necessitating a computation of beliefs by Bayesian updating given prior personal preferences. Alice or Barbara might perceive that the other prefers blue with some probability p and she prefers green by the remaining probability of $1 - p$. Depending on these probabilities as their beliefs in a strategic context of one-sided incomplete information, the equilibria become pair of strategies and belief conditions. Beliefs become more complex in conditions of two-sided incomplete information (Eichberger 1993; Friedman 1986).

It is possible that a player is uncertain about the other's intention to develop a common language in a VSG. She can misperceive the other as opposing the idea of discussing theoretical proposition via paintings. Suppose that Row assigns a subjective probability of $p(h)$ to the hypothesis h that Column is favorable to discuss structural constraints or dynamic nonunique anarchies via Rothko and Pollock. Row's belief $p(h)$ is not an objective event but reflects Row's psychological trait. Row may believe that Column's preferences do not exclude a visual semiotic communication yet in reality Column might have an opposite view. This can be exemplified by tossing a loaded dice implying *0* chance of landing *1* but one would believe that the dice is fair and therefore one would assign a probability of 1/6 to the event. Therefore, beliefs do not necessarily reflect objective states of the world.

The subjective probability of $p(r)$ measuring Row's belief that Column is hostile to the idea of making meanings through paintings, that is, Row's belief that Column is a "rejector," ranges between one and zero. Let "v" denote the event that Column is a "visual scholar," that is, Column conducts research in the visual turn in the Discipline. Row's belief is $p(r)$ before he receives the information about Column is updated upon Row's learning that Column is a visual scholar denoted by $p(r|v)$. By Bayes' theorem, Row's updated posterior belief $p(r|v)$, that is, Row's belief that Column is a rejector given that she is a visual turn scholar, becomes equal to $\dfrac{p(r).\ p(v|r)}{p(v)}$. The numerator is the multiplication of two probabilities, namely, Row's prior belief that Column is rejector $p(r)$ and the likelihood that given that Column is a visual scholar given that she is a rejector $p(v|r)$. The denominator is p(v), that is, the likelihood that Column is a visual scholar. For example, suppose that Row a priori thinks that Column is a rejector with a probability of *0.8*, that is, $p(r) = 0.8$. He then receives the information that Column might be a visual scholar, that is, $p(v) = 0.5$. If Row believes that Column is a visual scholar given that she is a rejector with probability *0.2*, then Row's posterior belief that Column is a rejector-given that she is visual turn scholar becomes $\dfrac{0.8.0.2}{0.5} = 0.32$. Hence, Row's prior-belief that Column is an objector with a likelihood of 0.8 reduces to 0.32 as Row receives information that Column is a visual scholar. The application of Bayesian updating in games of incomplete information implies insights along with such processes of reasoning.

We start with the assumption that Row is informed of his preference ordering quantified by his payoff function. He knows that he obtains positive payoffs only if *(A, A)* and *(R, R)* emerge as equilibrium outcomes out of the interaction in the Saussurean game; otherwise, he obtains nothing. Hence, Row's payoffs depend on the Column's strategy choices that is contingent on the Column's type that corresponds to Column's preference ordering. As to Column, she has two types due to the Row's misperception of her preferences. Suppose that the Row misperceives the Column as she supports to reject the idea of developing a sign altogether so that the Column chooses *R* regardless the Row does, that is, the Row believes that the Column has the dominant strategy of *R*. The Row thus believes that the Column obtains a utility of zero if she selects *A* regardless of the Row strategy choices. The Column is informed of the Row's suspicion about her, and the Row knows that the Column knows her suspicion and ad infinitum.[2] In other words, the Row's misperception is of common knowledge. What would the solution of such a game be?

---

[2] This condition is called Harsanyi Doctrine (Fig. 5.6).

Bayes-Nash equilibrium is the solution of such an incomplete-information game and therefore the answer of the question.

Suppose that Row believes that he interacts with Column in two alternative games with π and $1 - \pi$ probabilities, respectively. All pairs of uncoordinated choices generate zero payoffs for both players in both games. The top matrix below displays the game where both players get a payoff of 2 if they accept and a payoff of 1 if they reject that the painting constitutes a signifier. Column and Row possess no dominant strategies. Both have the incentives to reciprocate the other's choice of A and R. Column has the strictly dominant strategy R in the bottom game matrix. The two matrices below together model an asymmetric information Saussurean game in which Row is uncertain about Column's preferences, that is, types (Fig. 5.6).

The Row has one preference ordering in both game matrices as she has a unique type. The Column possesses two preference orderings and therefore two types that correspond to the Row's misperception of the Column. Let $T_1 = t$ denotes the Row's type and $T_2 = \{a, b\}$ where a denotes the Column's type in the upper game and b her type in the lower game. Type b has the dominant strategy of R.

The Cartesian product $\{t\} \times \{a, b\} = \{(t, a), (t, b)\}$ symbolizes possible type combinations. There is a function μ that maps type combinations into the probabilities corresponding to the Row's prior beliefs that she is in the upper and the lower games that are $\mu (t, a) = \pi$ and $\mu (t, b) = 1 - \pi$. The Row updates her beliefs about the Column's types given her knowledge that she is of type t. The Row's updated beliefs are her posterior beliefs of the Column's types. They are computed by Bayes' rule demonstrating that the prior beliefs of Row are equal to her posterior beliefs.

$$\mu'(a|t) = \frac{\mu(t,a)}{\mu(t,a) + \mu(t,b)} = \pi / (\pi + 1 - \pi) = \pi$$

$$\mu'(b|t) = \frac{\mu(t,b)}{\mu(t,b) + \mu(t,a)} = 1 - \pi / (1 - \pi + \pi) = 1 - \pi$$

			Column	
			A	R
		A	2, 2	0, 0
Row		R	0, 0	1, 1

$$[\pi]$$

			Column	
			A	R
		A	2, 0	0, 1
Row		R	0, 0	1, 2

$$[1 - \pi]$$

Fig. 5.6 Bayesian Saussurean Game

Players' strategies are type contingent. The Row has one type. She cannot have a strategy choice depending on her invariant type; her strategies remain A and R. The Column's strategies depend on her two types. The Column's type-contingent strategies are: "I agree on the sign regardless my preferences" denoted by (A, A); "I agree on the sign if I am of type a, but disagree if I am of type b" denoted by (A, R); "I disagree on the sign if I am of type a but agree if I am of type b" denoted by (R, A); "I disagree regardless of my type" denoted by (R, R). Type-contingent strategies of (A, A) and (R, A) are meaningless because type b is assumed to have the dominant action of R.

The next step is to compute the Row's expected payoffs to calculate her utilities deriving from A and R against the Column's type-contingent strategies of (A, A), (A, R), (R, A), and (R, R). The type b never chooses A because she has the dominant action of R, therefore (A, A) and (R, A) are eliminated. We are then left with (A, R) and (R, R).

The Row using her strategy A against (A, R) gets a utility of 2 with probability of π in the upper matrix where type a chooses A but a utility of 0 with probability of $1 - \pi$ in the lower matrix where type b chooses her dominant action of R. Hence, the Row obtains a utility of $2\pi + 0 (1 - \pi) = 2\pi$ from A against (A, R). The Row using her strategy R against (A, R) gets a utility of 0 with probability of π in the upper matrix where type a chooses A but a utility of 1 with probability of $1 - \pi$ in the lower matrix where type b chooses her dominant action of R. Hence, the Row obtains a utility of $0\pi + 1 (1 - \pi) = 1 - \pi$ from R against (R, R).

The Row using her strategy A against (R, R) gets a utility of 0 with probability of π in the upper matrix where type a chooses R and a utility of 0 with probability of $1 - \pi$ in the lower matrix where type b chooses her dominant action of R. Hence, the Row obtains a utility of $0\pi + 0 (1 - \pi) = 0$ from A against (R, R). The Row using her strategy R against (R, R) gets a utility of 1 with probability of π in the upper matrix where type a chooses R and a utility of 1 with probability of $1 - \pi$ in the lower matrix where type b chooses her dominant action of R. Hence, the Row obtains a utility of $1\pi + 1 (1 - \pi) = 1$ from R against (R, R).

The expected payoff table of the Row where the Row's strategies are displayed in the Columns and the Column's type-contingent strategies are displayed in the Rows is given below (Fig. 5.7):

There exist three candidates of strategy profiles: $\{R, (R, R)\}$, $\{A, (A, R)\}$, $\{R, (A, R)\}$. Starting with $\{R, (R, R)\}$, we observe that the Row obtains a payoff of 1 if she selects R but 0 if she selects A against (R, R). Thus, R is a best reply against (R, R) without depending upon any beliefs. Suppose that Row chooses R. Would either type a or b deviate from R given that the Row chooses R in the upper or lower

	A	R
AR	2π	$1 - \pi$
RR	0	1

Fig. 5.7 Row's expected payoff table

games? The answer is no: neither type would deviate from R to A against the Row's choice of R. The type a obtains a payoff of 1 from R against R but a payoff of 0 from A against R. Type b has the dominant action of R. Hence no player has any incentive to deviation under these conditions. Consequently, we obtain a Bayes-Nash equilibrium of $\{R, (R, R); $ *for all beliefs of the Row*$\}$.

The profile $\{A, (A, R)\}$ constitutes an equilibrium under a belief condition of the Row. We can read from the table that if $2\pi \geq 1 - \pi$, A constitutes a best reply of the Row against the Column's type-contingent strategy of (A, R). Thus, if $\pi \geq 1/3$, the Row chooses A against (A, R). Would type a deviate from A against the Row's choice of A? The answer is no. If she does, she obtains 0 instead of 2. Such a shift is quite costly for the type a. Type b has her dominant action of N. Therefore, no player would have any incentive to deviate from the proposed action profile. We obtain a second Bayes-Nash equilibrium: $\{A, (A, R); \pi \geq 1/3\}$.

The final possible equilibrium would contain $\{R, (A, R)\}$. However, there is no Bayes-Nash equilibrium in this case. Suppose that $\pi \leq 1/3$. As a result, the Row chooses R in both matrices. The type a obtains a payoff of 0 from A but she deviates to R, she obtains a payoff of 1. The type a has an incentive to deviate. Even though the type b has no incentive to shift from her dominant action, type a's deviation is enough to discard $\{R, (A, R)\}$ profile.

The two equilibria teach us under Row's misperception of Column two possible stable equilibria emerge, one that depends on Row's beliefs about Column and the other that does not. Row and Column select together R specifying a sure event of eliminating signifier quality of the painting. The rejection of both players does not depend upon Row's suspicion about Column. Nevertheless, the second equilibrium demonstrates how Row's uncertainty about Column's preferences plays a role is his choice of A under his belief that with at least one-third or higher possibility that Column selects A. Rubinstein's interpretations of mixed-strategy equilibrium are helpful in evaluations of both Bayes-Nash equilibria. The mixed-strategy equilibrium of $p = q = 1/3$ underlies Row's updated belief of π about Column's type. Both players' choice of the strategy A in the mixed-strategy equilibrium here turns into Row's belief of $1/3$ or greater that Column has no dominant strategy in R. The mixed strategy derives from both players' indifference, but the second Bayes-Nash equilibrium derives from the possibility of Row's choice of A provided Row believes that Column reciprocates Row's choice. Therefore, both equilibria are connected in this particular Saussurean game.

5.3   Bayes-Nash Equilibrium: Two-Sided Incomplete Information in Saussurean Games

Players' subjective beliefs about what each of them prefers about developing a semiotic channel of communication can be asymmetric in a mutual sense: both players can misperceive each other. Games of two-sided incomplete information are convenient tools to study implications of players' mutual misperceptions of the value of paintings' help in such an endeavor.

Suppose that both Row and Column possess two types such that the set of Row's types are denoted by r_1 and r_2 similar to Column's types denoted by c_1 and c_2. The types of r_2 and c_2 are assumed to have the dominant strategy of R. The types of r_1 and c_1 are assumed to have no dominant strategy. Accordingly, there are four type combinations: $(r_1 c_1), (r_1 c_2), (r_2 c_1), (r_2 c_2)$. The probabilities the Nature ascribes to pairs of types generate the distribution of prior probabilities (or priors). Priors are players' initial beliefs about each other's preferences the Nature selects (Osborne 2004, 502). They are denoted as $\mu = \{\mu (r_1 c_1), \mu (r_1 c_2), \mu (r_2 c_1), \mu (r_2 c_2)\}$ where $\mu (r_1 c_1)$ is read as the prior that Row and Column have no dominant strategy of R, $\mu (r_1 c_2)$ the prior that Row no dominant strategy but Column has the dominant strategy of R, $\mu (r_2 c_1)$ the prior that Row has the dominant strategy of R but Column has no dominant strategy, and $\mu (r_2 c_2)$ is the prior that both Row and Column have the dominant strategy of R. Let $p = \mu (r_1 c_1), q = \mu (r_1 c_2), r = \mu (r_2 c_1), s = \mu (r_2 c_2)$. Each pair describes a game and generates altogether a Saussurean game of two-sided incomplete information (Figs. 5.8, 5.9, 5.10 and 5.11):

Once Row and Column receive their types selected by Nature, they update their beliefs about each other and form their posterior beliefs.

By the Bayes' rule, Row's posterior belief of given that she has no dominant action, Column has no dominant action is:

$$\mu(c1 | r1) = \frac{\mu(r1,c1)}{\left(\mu(r1,c1) + \mu(r1,c2)\right)} = p / (p+q)$$

By the Bayes' rule, Row's posterior belief of given that she has no dominant action, Column has the dominant action of R is:

$$\mu(c2 | r1) = \frac{\mu(r1,c2)}{\left(\mu(r1,c2) + \mu(r1,c1)\right)} = q / (q+p)$$

By the Bayes' rule, Row's posterior belief of given that she has the dominant action of R, Column has no dominant action is:

$$\mu(c1 | r2) = \frac{\mu(r2,c1)}{\left(\mu(r2,c1) + \mu(r2,c2)\right)} = r / (r+s)$$

			Column	
			A	R
		A	2, 2	0, 0
	Row	R	0, 0	1, 1

Fig. 5.8 $[p = \mu (r_1 c_1)]$, Game 1

		Column	
		A	R
	A	2, 0	0, 1
Row	R	0, 0	1, 2

Fig. 5.9 $[q = \mu \, (r_1 \, c_2)]$, Game 2

		Column	
		A	R
	A	0, 2	0, 0
Row	R	1, 0	2, 1

Fig. 5.10 $[r = \mu \, (r_2 \, c_1)]$, Game 3

		Column	
		A	R
	A	0, 0	0, 1
Row	R	1, 0	2, 2

Fig. 5.11 $[s = \mu \, (r_2 \, c_2)]$, Game 4

By the Bayes' rule, Row's posterior belief of given that she has the dominant action of R, Column has the dominant action of R is:

$$\mu(c2 \,|\, r2) = \frac{\mu(r2,c2)}{\left(\mu(r2,c2) + \mu(r2,c1)\right)} = s/(s+r)$$

Similarly, by the Bayes' rule, Column's posterior belief of given that she has no dominant action, Row has no dominant action is:

$$\mu(r1 \,|\, c1) = \frac{\mu(r1,c1)}{\left(\mu(r1,c1) + \mu(r2,c1)\right)} = p/(p+r)$$

By the Bayes' rule, Column's posterior belief of given that she has no dominant action, Row has the dominant action of R is:

$$\mu(r2 \,|\, c1) = \frac{\mu(r2,c1)}{\mu(r2,c1) + (\mu(r1,c1)} = r/(r+p)$$

By the Bayes' rule, Column's posterior belief of given that she has the dominant action of R, Row has no dominant action is:

$$\mu(r1 \,|\, c2) = \frac{\mu(r1,c2)}{\mu(r1,c2) + (\mu(r2,c1)} = q/(q+s)$$

By the Bayes' rule, Column's posterior belief of given that she has the dominant action of R, Row as well has the dominant action of R is:

$$\mu(r2\,|\,c2)=\frac{\mu(r2,c2)}{\mu(r2,c2)+(\mu(r1,c2)}=q/(q+s)$$

As strategies in games of incomplete information are type-contingent, we remark that two actions associated with two types generate four type-contingent strategies for Row and Column. They are AA, AR, RA, RR for each player. The strategies of AA and RA are eliminated because they prescribe the dominated action of A for the types r_2 and c_2. As a result, we are left with the strategies of AR and RR read respectively as "if I am type 1, I select A and if I am type 2, I select R" and "I choose R regardless my type" for both Row and Column. Consequently, there exist four combinations which are considered as candidates of Bayes-Nash equilibrium action profiles: AR against AR, AA against RR, RR against AR, and RR against RR.

We start with Row's expected payoffs in each of the four combinations associated with her beliefs:

1. *Row's payoff to AR against AR*

   Row's *type r_1* chooses A and type r_2 chooses R while Column's type c_1 chooses A; type c_2 chooses R:

$$2\left(\frac{p}{p+q}\right)+0\left(\frac{q}{q+p}\right)+1\left(\frac{r}{r+s}\right)+2\left(\frac{s}{s+r}\right)=2\left(\frac{p}{p+q}\right)+\left(\frac{r+2s}{r+s}\right)$$

2. *Row's payoff to AR against RR*

   Row's type r_1 chooses A and type r_2 chooses A while Column's types c_1 and c_2 choose R:

$$0\left(\frac{p}{p+q}\right)+0\left(\frac{q}{q+p}\right)+2\left(\frac{r}{r+s}\right)+2\left(\frac{s}{s+r}\right)=2$$

$$0\big(p/(p+q)\big)+0\big(q/(q+p)\big)+2\big((r/(r+s)\big)+2\big((s/(s+r)\big)=2$$

3. *Row's payoff to RR against AR*

   Row's type r_1 and type r_2 choose R while Column's type c_1 chooses A and c_2 choose R:

$$0\left(\frac{p}{p+q}\right)+1\left(\frac{q}{q+p}\right)+1\left(\frac{r}{r+s}\right)+2\left(\frac{s}{s+r}\right)=\left(\frac{q}{q+p}\right)+\left(\frac{r+2s}{s+r}\right)$$

4. *Row's payoff to RR against RR*

   Row's type r_1 and type r_2 and Column's type c_1 and c_2 choose R:

$$1\left(\frac{p}{p+q}\right)+1\left(\frac{q}{q+p}\right)+2\left(\frac{r}{r+s}\right)+2\left(\frac{s}{s+r}\right)=\left(\frac{q}{q+p}\right)+\left(\frac{r+2s}{s+r}\right)=3$$

We now turn to Column's expected payoffs in each of the four combinations associated with her beliefs:

5. *Column's payoff to AR against AR*
    Column's type c_1 chooses A; Column's type c_2 chooses R while Row's type r_1 chooses A and type r_2 chooses R:

$$2\left(\frac{p}{p+r}\right)+0\left(\frac{r}{r+p}\right)+1\left(\frac{q}{q+s}\right)+2\left(\frac{s}{s+q}\right)=2\left(\frac{p}{p+r}\right)+\left(\frac{q+2s}{q+s}\right)$$

6. *Column's payoff to AR against RR*
    Column's type c_1 chooses A; Column's type c_2 chooses R while Row's type r_1 and type r_2 choose R:

$$0\left(\frac{p}{p+r}\right)+0\left(\frac{r}{r+p}\right)+2\left(\frac{q}{q+s}\right)+2\left(\frac{s}{s+q}\right)=2$$

7. *Column's payoff to RR against AR*
    Column's type $c1$ and $c2$ choose R while Row's type $r1$ chooses A and Row's type $r2$ chooses R:

$$0\left(\frac{p}{p+r}\right)+1\left(\frac{r}{r+p}\right)+1\left(\frac{q}{q+s}\right)+2\left(\frac{s}{s+q}\right)=\left(\frac{r}{p+r}\right)+\left(\frac{q+2s}{q+s}\right)$$

8. *Column's payoff to RR against RR*
    Row's type r_1 and type r_2 and Column's type c_1 and c_2 chooses choose R:
    $1(p/(p + r)) + 2(q/(q + s)) + 1 ((r/(r + p)) + 2(s/(s + q)) = 3$
Players' expected payoffs resulting from their type-contingent strategies are tabulated in the table below (Fig. 5.12):
Figure 5.18 allows the computation of Bayes-Nash equilibria:

		Column	
		AR	RR
Row	AR	$2\left(\frac{p}{p+q}\right)+\left(\frac{r+2s}{r+s}\right)$, $2\left(\frac{p}{p+r}\right)+\left(\frac{q+2s}{q+s}\right)$	$2, \left(\frac{r}{p+r}\right)+\left(\frac{q+2s}{q+s}\right)$
	RR	$\left(\frac{q}{q+p}\right)+\left(\frac{r+2s}{s+r}\right), 2$	$3, 3$

Fig. 5.12 Expected payoff table of row and column

(AR,   AR):   *AR*   is   a   best   reply   against   *AR*   for   Row   if

$$2\left(\frac{p}{p+q}\right)+\left(\frac{r+2s}{r+s}\right)>\left(\frac{q}{q+p}\right)+\left(\frac{r+2s}{s+r}\right)$$, therefore, if $p \geq \frac{1}{2} q$. Assume that $p \geq \frac{1}{2}$

q. AR is a best reply against AR for Column if $2\left(\frac{p}{p+r}\right)+\left(\frac{q+2s}{q+s}\right)>\left(\frac{r}{p+r}\right)+\left(\frac{q+2s}{q+s}\right)$

, therefore if $p \geq \frac{1}{2} r$. Assume that both belief conditions hold. Consequently, {*(AR,*

AR); $p \geq \frac{1}{2} q$ *and* $p \geq \frac{1}{2} r$} constitute a Bayes-Nash equilibrium.

(RR, RR): *RR* is a best reply to *RR* as $3 > 2$ for any beliefs. Thus, the second equilibrium is independent of players' beliefs. Consequently, strategy-belief pair of {*(RR, RR)*; *for all beliefs*} constitutes an equilibrium.

For Column, her best reply against Row's *RR* strategy is not *AR* but *RR* as $2 < 3$. Thus, *(RR, AR)* pair cannot be part of any Bayes-Nash equilibrium. Similarly, Row has an incentive to deviate from *AR* to *RR* to obtain *3* instead of *2* against Column's *RR* strategy. Thus, *(AR, RR)* pair cannot be part of any Bayes-Nash equilibrium.

The equilibria imply that if Row and Column perceive each other as being fully opposed to the idea of developing a visual semiotic communication channel with paintings functioning as signifiers, their mutual beliefs about each other's preferences do not matter; they always reject. The interaction itself is based upon the view that players are intentional beings and able to attribute mental positions such as their openness to the idea that they can learn more about structural constraints and dynamic anarchies by interpreting abstract art. The equilibrium of {*(RR, RR)*; *for all beliefs*} demonstrates that when players interact with one another under their beliefs of the other as a staunch opponent, find optimality of coordination in mutual rejection, not in mutual agreement. Thus, a type of the player who does not desire to oppose a visual coordination behaves in the direction of the refusal if she believes that the other player shares the same opinion. Players' beliefs span both internal and external states of mind, thus constitutes mental reasons of strategic choices in games of interactive rationality. Once they perceive that each has an incentive to refuse to coordinate on the development of a semiotic language, then their beliefs do not matter. Mutual rejection becomes optimal.

The equilibrium of {*(AR, AR)*; $p \geq \frac{1}{2} q$ *and* $p \geq \frac{1}{2} r$} reflects reciprocity of choices so that both can accept or reject to communicate on the basis of abstract paintings of Pollock and Rothko. Naturally, types with dominant actions of rejection are not amenable to change their minds. Yet the equilibrium indicates that a reciprocity of choices can occur. Players can coordinate in direction of mutual agreement without depending on $s = \mu (r_2 c_2)$. The belief condition excludes $s = \mu (r_2 c_2)$ but implies that $p \geq \frac{1}{2} q$ and $p \geq \frac{1}{2} r$, that is, their priors of being in game 1 exceed or at least equal to their priors of being involved in games 2 and 3. Game 1 attaches a higher value to the coordination on the *(A, A)* outcome than the *(R, R)* outcome. Game 2 and game 3 imply the dominance of rejection either on the side or Row or Column. Therefore, players' mental positions about one's inclination to reject as the dominant strategy must be sufficiently small for the first equilibrium to realize. While there is no chance for players' *(RR, RR)* payoff of *3* to be beaten with alternative choices, it is instructive to examine how beliefs affect players' equilibrium payoffs in the game.

5.4   Repeated Interactions

Repetitions of interactions take form through repeated games and evolutionary games. A Saussurean or a Peircian game can be repeated finitely so that players know or not when their interaction ends. If they know when their interaction ends, like for example, they play the same stage game twice or for any finite number, then they play a finitely-repeated game; otherwise they are involved in an infinitely-repeated game. Without going into the formal analysis of such games, it must be noted that, for example, if a Saussurean game is repeated finitely many times, either *(A, A)* or *(R, R)* or any combination of these equilibria are obtained. No other possible equilibria exist as unilateral rejection and acceptance generate zero payoffs to players. If, for example, the dress color game is repeated finitely many times, then there can be only four types of equilibrium, as there exist two Nash equilibria and no outcome exists that is not an equilibrium but that benefits both Alice and Barbara (Watson 2008, 259–261). It follows that the equilibria of the finitely repeated dress color game consist of (Blue, Green) or (Green, Blue) played each period or any other combination of these equilibria throughout rounds of interaction. Thus, Alice and Barbara will never end up in wearing dresses in red or yellow if there are repetitive parties over time. Alice always wears blue, and Barbara always wears green or vice versa or they alternate their dress colors in parties to come.

In evolutionary terms, one can assume, for example, that dress colors that are successful contribute to the fitness of Alice and Barbara in their ongoing social activities (Maynard Smith 1982; Sigmund 1993; Samuelson 1997; Hofbauer and Sigmund 1998; McElrath and Boyd 2007). An evolutionary game would approach the dress color selection for a party considering a large population of party goers selecting between blue, green, red, and yellow. Given a population where each of these different colors are adopted, perhaps blue becomes selected more often over time as it is successful, that is, blue wearers' fitness increases compared to other colors in finding partners, achieving marriages, and having children. Hence, the evolutionary game does no longer focus on individual players but on colors. Yet, colors are not rational agents who try to get their best payoffs in a game. They simply indicate proportions of the populations. We forget about Alice and Barbara being rational trying to satisfy their preferences in the game. Colors interact, not individuals. Some colors survive and some not in an evolutionary setting. Suppose that partygoers are mixed wearing green, red, and yellow. Now suppose blue enters the population, that is, some people wear blue dresses in parties. The newcomer color is qualified as a "mutant." If, given the proportions of colors other than blue, the mutant attracts more partners and the fitness it provides increases compared to green, red, and yellow, blue, then it will ultimately invade the population. Thus, the initial color mixture of green, red, yellow is not evolutionarily stable but blue becomes the evolutionarily stable color over time. The evolutionarily stable equilibrium indicates that no one will wear green, red, or yellow dresses. The color blue generating a higher fitness will be selected more often. And everyone will ultimately wear blue. Now we turn to evolutionary Saussurean games.

5.5   Evolutionary Equilibrium in Saussurean Games

Evolutionary equilibrium is meaningful when a number of IR scholars interact over paintings and paintings' meaning for theoretical propositions. Interactions among scholars imply regularities reached over time constituting conventions about what sign an IR proposition and a painting together form. An IR scholar opposes the convention and rejecting the iconicity of the painting, then she might set an example of a rejector (a mutant) in the population of IR scholars. She might be imitated by others and destabilizing iconness of the painting depending on the size of the majority following the convention. VSGs formulated as evolutionary games examine conditions of the stability of conventions.

Evolutionary games offer a perspective greatly enriching the view of dynamic nonunique anarchies of Wendt (1992). "Some states for whatever reason may become predisposed toward aggression. The aggressive behavior of these predators or 'bad apples' forces other states to engage in competitive power politics, to meet fire with fire, since failure to do so may degrade or destroy them" (Wendt 1992, 408). States emulate successful aggressive policies in a competitive anarchy of self-help. States emulate successful altruist policies in a cooperative security system reflecting the exact opposite of competitive anarchies. States emulate policies of indifference toward others yet being attentive to issues that may endanger their security in an anarchy Wendt calls "individualistic" between the two extreme ones (Wendt 1992, 400). Why emulation? The answer is simple: emulation make states successful; maximize security in a context of a behavioral sameness ruling systems they coexist. Anarchies as social systems constrain states' interactions in accordance with prevailing norms they generate.

Evolutionary Saussurean games shift attention from interacting states to interacting players who generate signs. Two major axioms of evolutionary games concern rationality and ongoing interactions. Evolutionary games constitute a sharp turn from the rationality assumption and one-shot games. They are dynamic, so that they model on-going interactions among actors and also how strategies change through time. The game allows players to make mistakes as well; players are not perfectly rational individuals and learn to select better strategies progressively. Players can become more resourceful being creative and inventive in communications of theoretical interpretations. They become successful by proposing meanings of theoretical sentences on the basis of abstract paintings and by creating a new theoretical language. They may learn that agreeing or not on a painting as a signifier is a better strategy. The strategies of accept and reject become modes of behavior. Players learn to imitate those who are successful in communicating with others. Hence, an evolutionarily stable equilibrium (ESS) of (A, A) implies that an abstract painting by Pollock or Rothko emerges in Peircian terms as a symbol of dynamic anarchies or structural constraints through time (perhaps decades). Players' coordination on a painting implies the evolutionary stability of the (A, A) or the (R, R) outcome. Therefore, evolutionary games constitute a powerful tool of equilibrium selection in Saussurean games. They answer quite easily the question of whether players agree or disagree on the painting's signifier quality.

The definition of evolutionarily stable strategy by Maynard Smith (1982, 14) implies that:

A is an ESS if:

1. *either E (A, A) > E (R, A)*
2. *or E (A, A) = E (R, A) and E (A, R) > E (R, R)*

Condition 1 implies that A fares strictly better than R against itself, that is, no player using R survives and gets extinct in a population where everyone uses A. Hence, the rejection of sign relation generates a lesser success in a population consisting purely of IR scholars who insist that Rothko painting (or a Pollock painting) is a signifier and Waltz's structural constraints (Wendt's dynamic anarchies) is the signified. The acceptance of the sign relation is then an evolutionarily stable strategy. Condition 2 takes a possibility that the rejection of the sign relation is equally successful as acceptance in a population of IR scholars who support the sign relation. In this case, the acceptance of the sign relation must satisfy an additional condition: it must be successful in a population of IR scholars who reject the sign relation, so that Y can invade a population purely consisting of the rejectors. If Y satisfies either conditions, then it is an ESS.

The evolutionary approach to game 1 reproduced below is instructive in how to compute an ESS (Fig. 5.13):

We assume that there exists a large population of players (IR scholars) who interact pairwise. Every player normally agrees that, say, Rothko painting is a signifier of structural constraints. Therefore, those players who agree and select strategy A obtain a payoff of $a > 0$ in encountering with each other. Otherwise, if they encounter those who reject, then they obtain a payoff of 0. Similarly, those who reject obtain $b > 0$ in encounters with each other; otherwise they obtain a payoff of 0.

Now suppose that players accept that the Rothko painting constitutes a signifier of structural constraints but a small fraction ε of players who select R called "mutants" enter the population. Mutants obtain a payoff of $b > 0$ in their encounters with each other; they obtain 0 otherwise. Suppose that the population of players is large so that the fraction of mutants in the entire population of players equals the fraction of mutants in the remaining portion of the population (Osborne and Rubinstein 1994, 48; Osborne 2004, 394). Hence, a player who selects A encounters a mutant using R with probability ε and encounters a player using A with $1 - \varepsilon$ probability.

Call players who agree on the sign relationship as acceptors. Acceptors' expected payoff becomes: $a (1 - \varepsilon) + 0 (\varepsilon) = a (1 - \varepsilon)$. Call mutants who reject the sign relationship as rejectors. Rejectors' expected payoff becomes $0 (1 - \varepsilon) + b(\varepsilon) = b\varepsilon$.

		Column	
		A	R
	A	a, a	0, 0
Row	R	0, 0	b, b

Fig. 5.13 Game 1

If acceptors' payoff exceeds rejectors' payoff, that is, if $a (1 - \varepsilon) > b\varepsilon$, and, therefore, $\varepsilon < a/(a + b)$, A is evolutionarily stable. Thus, an IR scholar in a population of adherents finds that supporting the sign relation is more beneficial, and, as a result, paintings become symbols of structural constraints or dynamic anarchies ultimately. If players' usual reaction to visual sign formation is summarized by the strategy R, hence the population is composed by rejectors, then mutant acceptors entering and constituting a fraction ε of the population, the evolutionary stability of R depends on the condition of the ε value. Rejectors' payoff is $0 (\varepsilon) + b (1 - \varepsilon) = b(1 - \varepsilon)$. Mutant acceptors' payoff is $a (\varepsilon) + 0 (1 - \varepsilon) = \varepsilon a$. Therefore, if $b(1 - \varepsilon) > \varepsilon a$, that is, if $\varepsilon < b/(a + b)$, R is evolutionarily stable.

Thus, in general, if the population normally accepts or rejects the painting as a signifier and if mutants are more successful than adherents to the generally accepted norm, the critical threshold of ε decreases. In the case of rejectors constituting mutants, if mutants' payoff increases, then acceptors become vulnerable to invasions of rejectors. This is similar to the case of a population normally rejects and mutants are acceptors. The population becomes more vulnerable to mutant behavior of acceptors who reap higher payoffs compared to rejectors. The population will fully consist of acceptors over time.

In general, suppose that some players start to reject the meaning of the Rothko painting as signifying structural constraints in a large population of acceptors who successfully communicate by agreeing on the sign. The entrance of rejectors in the population means the emergence of a cultural opposition. The opposition fails provided that the number of mutant rejectors is below a critical threshold depending on rejectors' payoff. Hence, if the population consists of mostly rejectors, then the fraction of acceptors must be below a smaller threshold for the rejection of the sign to be an ESS. Acceptors or rejectors as mutants can invade populations depending on how successful they are in encounters.

Moreover, there is no need for high R payoffs for R to be ESS in a population of rejectors (or by the same token no need for high A payoffs); it is enough that (R, R) (or (A, A)) is a strict Nash equilibrium. Only the resistance to mutant invasions varies. If, for example, rejectors obtain a lesser success among themselves compared to the success of acceptors in populations of acceptors, acceptors constituting smaller fractions can invade rejector populations. The evolution of A and R, and, therefore, paintings that become symbols over time, is highly dependent upon the success they yield.

It is impossible that the paintings convey no ideas, no interpretation grounds. The equilibrium or equilibria of the game generate nonempty intersections of interpretations. Players become involved in a process of interpretation liberating them from linguistic mazes of theoretical arguments. The artwork then becomes nonlinguistic sources to interpret the main propositions of Structural Realism and Constructivism. Thus, the game becomes a context to destabilize known arguments focusing on anarchy conceptualizations by Kenneth Waltz and Alexander Wendt. Peircian games enrich the hybrid method by taking three distinct signs into account.

Chapter 6
Peircian Games

In Saussurean games, Row entertains beliefs about what beliefs Column has concerning Row's strategies of accept and reject, and, similarly, Column entertains beliefs about what beliefs Row has concerning Column's strategies of accept and reject. In Peircian games these coupled subjective beliefs are about three strategies, namely, icon, index, and symbol, instead of two. Players are assumed to know the meanings of icon, index, and symbol. They have figured out the three coordination equilibria each connected with the three distinct signs. Thus, Peircian games pose a greater coordination challenge compared to Saussurean ones.

6.1 Peircian Games in Strategic Form

Peircian games allow players to select among three strategies: propose icon, index, or symbol as sign relations. Consequently, Peircian games do not permit players to reject signs unlike Saussurean games.

A generic Peircian game of complete-information game in strategic form is given in Fig. 6.1. Players interact by proposing that a Rothko or a Pollock painting either constitutes an icon or an index or a symbol for Waltz's structural constraints or Wendt's nonunique dynamic anarchies, respectively.

Saussurean signifier-signified sign relation is embedded in Peircian semiotics: Charles Sanders Peirce's semiotics approach offers the third element. Peirce asserts that "a sign, or representamen, is something which stands to somebody for something in some respect or capacity. It addresses somebody, that is, creates in the mind of that person an equivalent sign, or perhaps a more developed sign. That sign which it creates I call the interpretant of the first sign. The sign stands for something, its object. It stands for that object, not in all respects, but in reference to a sort of idea, which I have sometimes called the ground of the representamen" (Peirce 1955, 99). Hence, Peirce discusses three elements: the representamen which is perceptible as the signifier, the interpretant which is the mental image the recipient forms of the

© Springer Nature Switzerland AG 2023
S. Ş. Güner, *Art and IR Theory*, Mathematics in Mind,
https://doi.org/10.1007/978-3-031-32342-3_6

		Column		
		Icon (I)	Index (X)	Symbol (S)
Row	Icon (I)	a, b	0, 0	0, 0
	Index (X)	0, 0	c, d	0, 0
	Symbol (S)	0, 0	0, 0	e, f

Fig. 6.1 Generic Peircian game

sign as the signified, and the object which refers to something beyond the sign (Nöth 1990, 42–47).

Each strategy forms a signification process consisting of three elements: representamen (or sign), interpretant, and object. To illustrate, take a Pollock drip-paint work. Wendt's conceptualization of dynamic multiple anarchies is the object. Wendt's idea is the representamen. The sign is the painting. It stands for Wendt's conceptualization. If the player thinks that the painting exactly corresponds to Wendt's conceptualization, then it is an icon. If a player thinks that the painting corresponds to the evidence, the existence of idea of the multiplicity of dynamic anarchies, then the sign becomes an index. If a player thinks that the painting is to be established as a convention in the Discipline over time for Wendt's conceptualization, then the painting is a symbol. The interpretant is what each sign produces in the mind of the player. It can be about the implications of selecting a strategy and coordinating with other players' selections for the Discipline. What would, for example, mean for an individual player the achievement of a common language based on players' coordination of icon, index, or symbol? Answers to the question indicate alternative interpretants. Hence, the truth is intersubjective; it pertains to strategic interaction and uncertainty the game represents. A VSG describes how mental subjectivities get intertwined in the strategic interaction delimited by the implications of three signs.

6.2   Icon

An icon produces meaning from physical, identical resemblance. It bears a resemblance to its object in some way: it looks or sounds like it (Fiske 2011, 44–45). If, for example, a player selects a Rothko painting as an icon for Structural Realism, she means that the painting corresponds to structural constraints exactly (Waltz 1979, 40). Hence, in the context of meaning making for Waltz's structural constraints by the help of a Rothko painting, icon means one-to-one correspondence between the painting and the graph. Thus, a Rothko painting can be selected as an icon of Waltz's figure. The painting enables a player to establish a correspondence of mental identity between the figure and the painting. The iconness of a painting cannot be made true or false by some worldly facts independent of players' minds.

Similarly, a player who claims that there exists iconicity relation between a Pollock drip-painting, like Blue Lines, then she means that the painting physically resembles Wendt's conceptualization of anarchy; an identity relation exists between the painting and the conceptualization. She interprets the painting's swirls and curls in different colors as they physically resemble different anarchies evolving. The claim of icon, in this context, "signifies by resembling or by sharing some quality with its object," that is, the painting is interpreted as resembling to or sharing the dynamic quality of Wendt's nonunique anarchy conceptualization. There is a likeness between the painting and the proposition a player constructs in her mind. The mental construction of likeness generates an iconic sign, an iconicity of the Pollock painting regarding Wendt. The other, however, may believe that the sign of the painting is not an icon but an index or a symbol. She might believe that the painting does not exactly correspond to Wendt's anarchy conceptualization; no firstness relation exists between Pollock and Wendt but a secondness or perhaps a thirdness. Firstness refers to icon, secondness to index, and thirdness to symbol. Concerning Peirce's concepts of icon, index, and symbol Elkins (2003, 12) indicates that "A First is something in itself, a Second exists in dyadic relation to another thing, and a Third is a mean that cannot be separated from a law or a purpose." Therefore, firstness implies that Pollock's work mirrors Wendt's anarchy proposition, secondness implies that Pollock's work relates to Wendt's anarchy proposition, thirdness implies that Pollock's work is to be established by convention among IR scholars as a symbol of Wendt's anarchy proposition; Pollock-Wendt connection is arbitrary. The player insisting on the indexity of Pollock for Wendt bases her argument that the painting affects her thought. There is a physical or a causal connection between Pollock and Wendt formed in her mind. Thus, while icon refers to a direct correspondence an index remains as an indirect correspondence as Pollock's artwork affects player's thought and intuition in the direction of the complexity of Wendt's arguments. Thus, increments in the abstraction level make the sign from icon to index and index to symbol (Mittelberg 2006, 40).

The player's proposition that Pollock versus Wendt corresponds to icon is equivalent to propose a firstness relation. The other player who does not deny that the firstness exists can propose a secondness, that is, she claims that the correspondence between the painting and the proposition takes the form of a dyadic relation: the artwork corresponds to the idea of nonunique dynamic anarchies. Iversen (2018, 1) indicates that Peirce evaluated "*index* as the most 'forceful' type of sign: it signifies by establishing an existential or causal link to its referent, either by directing our attention to something or by being physically impressed or affected by it."[1] Thus, the secondness establishes a causal or an existential link between Pollock and Wendt so that the features of the painting do not mean a causal but an existential link. There is no causal link as if Pollock's work generated Wendt's proposition (not an inverse causal relation contradicting the timeline). Twists and curls of paint in a Pollock painting direct players' attention to Wendt's anarchy conceptualization. They demonstrate the idea of the multiplicity and dynamic character of anarchies in the absence of which a player could not infer the painting stands for the anarchy

---

[1] Italics in the original.

concept in terms of secondness. Therefore, there is no direct causal connection between Wendt and Pollock according to the index argument of the player. However, a causal connection between the painting and Pollock's bodily movements, handling paint, and his hands flowing on the surface exists: Pollock's drip paintings are clearly indexical – think of how *Out of the Web* reveals the movement of Pollock's body, his physical presence as he made the painting" (Hatt and Klonk 2006, 209). Thus, the other player's index proposition is based on an ideational correspondence, not a causal one between the painting and the proposition. Hence, a player who is convinced of the dynamic nonunique quality of anarchy might see it in a Pollock painting saying "there!" (Peirce 1960a, b, c, 211).

Some features in a Pollock painting correspond to the player's mental interpretation of stands for a player. However, why would IR theorists feel dynamic nature of anarchy by gazing a Pollock painting but not one created by Mark Rothko? Does a curl, a swift directional change in drip-painting elevate a Pollock painting to the rank of an icon in constructivist universe? We know that a linear line with a positive or a negative slope as well indicates change; a curl does not necessarily possess a meaning monopoly to subsume change. The features as signs constitute a tripartite entity as Peircian semiotics take signs as a combination of "the sign itself, the thing signified and the cognition or feeling produced in the mind of the interpreter" (Iversen 2018, 1). Peircian semiotics thus allow one to concentrate on every color, curl, turn in a Pollock drip-paint work and placement, hue, and nature of colors in a color-field painting by Rothko according to Peirce's sign concept as a tripartite entity. The result will be a multiplication of signs each aspect generates in the minds of players. The multiplicity then will allow further interpretations of propositions offered by Waltz and Wendt. Players' talking and exchanging their feelings before they select strategies constitute a pre-game phase in VSGs.

Players' talking to one another constitutes a channel of communication revealing lines of contention and accord among IR theorists. The type of reasoning leading to signifying constructs is central (Fiske, 2011, 1). A successful communication will guide and help scholars to put facts together so that they obtain an appreciation of mutual arguments. Players then can appreciate that no sharp distinction or boundary between sign propositions. Both players may agree that a Rothko painting does not eliminate dynamic or changing anarchies as colors interact in it. The intimate feeling of color interaction in a Rothko painting for a player can be shared by the other leaving artists' intimate feelings aside. Players are not obliged to search for Pollock's and Rothko's feelings about what their artwork express for themselves and compare them with their own or both artists' view about what their canvases represent. Any comparison of these sorts does not bear any clue about how the anarchy concept is discussed in Structural Realism and Constructivism. What counts is IR scholars' sign-making efforts facilitate or not their coordination. The meanings of signs help players to classify how they make meaning out of the paintings and to reduce the range of possible excitement, emotions, and feelings. In short, icon, index, and symbol put an order over players' exchanges in, for example, before they actually select strategies.

Those IR scholars admitting the iconicity of Pollock painting are qualified as "strict types" who claim that the sign relation must be based upon a nonarbitrary strict resemblance relation. The "non-strict types" in turn reject iconicity admitting that a Pollock painting constitutes an index or a symbol of Wendtian anarchy conceptualization.

6.3 Index

Index is a sign that has a direct existential relationship to its object, in terms of a physical or a causal bound. There is a direct link between the object and the sign; the two are actually connected. The smoke is an index of fire, for example. Unlike an icon, it does not imply any identical correspondence between itself and its object/ referent. It produces meaning by revealing some connection at an ideational level similar to smoke indicating fire. Also, an index differs from icon, because it is not based upon a relationship of reflexivity or resemblance. If a player selects index concerning sign making in Rothko-Waltz connection, then it means that she believes a Rothko painting supports the idea that structural constraints exist and are present, so that structures of international systems shape and shove states' interactions.

The paintings by Pollock and Rothko taken as indices do not correspond to arguments of Wendt and Waltz, respectively but they direct our attention to these arguments. A Rothko painting indicates structural constraints not through direct resemblance like an icon but through a perception of color interactions at ideational level. The painting and Waltz's structural constraints are not directly or actually connected. A Pollock painting constituting an index of Wendtian anarchy concept not because there is a physical or causal connection between the painting and the concept but on the basis of a mental connection. An IR theorist can contend that a Pollock painting can mentally indicate Wendt's anarchy concept. Thus, icon and index connect to their objects differently.

Wendtian anarchy concept and a Pollock drip-paint work. Hence, she claims that Pollock's painting corresponds to and reflects Wendt's anarchy conceptualization. Pollock's painting is therefore an icon in Peircian terms. Icon, index, and symbol denote distinct signs and therefore semiotic modes in Peircian terms. Yet, in terms of purity they mix with each other through porous mental borders. Icon, index, and symbol are not mutually exclusive categories. They are semiotic modes that can overlap in a particular process of signification: "There are thus in principle no pure icons, indices, or symbols, and that a sign can represent any combination of the three. These modes of relationships between the sign vehicle and its referent reflect the degree to which the signified determines the signifier. While iconic and indexical signs tend to be highly motivated through similarity and contiguity respectively by their referent, their interpretation can still rely on cultural conventions (such as in the case of films and photographs). Symbolic signs, on the other hand, tend to exhibit a high degree of arbitrariness and need to be learned. But they can also incorporate iconic dimensions, as it is the case with onomatopoeic words or sign language signs" (Mittelberg 2006, 37).

Players can choose among them to propose alternative meanings a painting generates for an IR proposition. The strategies stem from players' interpretations of shapes, colors, and their placements in a painting conveying feelings and meanings. Hence, each sign corresponds to a strategy.

Peirce indicates that a sign is by its nature triadic: a sign is "something which stands to somebody for something in some respect or capacity" (Elkins 2003, 12). Peirce defines a sign as "something which stands to somebody for something in some respect or capacity. It addresses somebody, that is, creates in the mind of the person an equivalent sign, or perhaps a more developed sign. That sign which it creates I call the interpretant of the first sign. The sign stands for something, its object. It stands for that object, not in all respects, but in reference to a sort of idea, which I have sometimes called the ground of the representamen" (Peirce 1955, 99). Peirce's definition implies, for example, a player takes a Rothko painting as a sign of Structural Realism. The painting is the representamen: it stands for a structural realist claim for the player in question. The interpretant is a more developed sign than the painting. It is in the mind of the player who interprets the painting as containing a trait corresponding to a specific structural realist claim, say about structural constraints. At the final step, the interpretant paves the way for the painting to represent Structural Realism as a dynamic theory and becomes the ground of the representamen, that is, the object. It follows that the interpretant affirms that Constructivism is a dynamic IR theory. The object, that is, the ground of the representamen, implies more than three evolving anarchies. Therefore, the object generates a meaning that goes beyond Hobbesian, Lockean, and Kantian anarchies.

Pollock's Blue Lines and the interpretation of the blue lines painted in the artwork as corresponding to anarchies should emanate from mental sources. First, lines are interpreted like alternative anarchies. The multiplicity of the lines refers to the multiplicity of possible anarchies in international systems. Once a resemblance relation is set, and the lines become icons of different anarchies. The changing angles of these lines in the canvas constitute an alternative iconicity implying a distinctness relation. Floch (1985, 136) examines Kandinsky's "Composition IV" piece by piece: "Le tableau semble donc s'articuler en deux parties, droite et gauche, qui s'opposent et selon une saisie paradigmatique (la reconnaissance d'une coprésence dans le tableau des termes opposés de quelques catégories visuelles constitue la projection du paradigmatique sur l'axe syntagmatique; une telle organisation contrastive permet ainsi de parler de structure textuelle) et selon une saisie syntagmatique (la mise en relation des formes est toujours binaire mais elle se fait par une double reprise à droite et par deux symétries à gauche)." Thus, it is possible to interpret Pollock's "Blue Lines" to detect alternative signs of anarchies Wendt proposes. The whole painting or portions of it permit iconicity interpretation. The drip-paint swirls in the painting meaning identical resemblance to alternative and evolving anarchies constitute an iconicity relation different than blue lines. VSGs concentrate on players' strategic interdependence in the context of iconicity so that players' preferences can vary. Players would not prefer to interpret "Blue Lines" but trajectories of dripped paint as standing for changing multiple anarchies as icons. They might also prefer to interpret blue lines or trajectories of dripped paint as forming an index or a symbol corresponding to alternative anarchies. The strategic interdependence covering all three signs displays the complexity of sign making.

The strategy of color for example incorporates a sensation about how it represents a particular theory. Two interacting scholars making meaning of a Rothko, or a Pollock painting must have solved the problem of why contemplating them makes sense or not in interpreting Structural Realism or Constructivism. According to Schapiro (1972–1973, 12) "Juan Gris remarked that a patch of yellow has a different visual weight in the upper and the lower parts of the same field." Crytesthesia is the result of such organizational changes (ibid.). Thus, the organization of colors generates different visual weights depending on how colors are placed in the painting.

6.4   Symbol

The selection of symbol means that the player rather believes that Rothko's painting (or all color-field paintings by Rothko) will form a convention in the Discipline; all scholars will ultimately accept the painting as a referring to Waltz's proposition. There exists no connection or resemblance between paintings and theoretical statements in symbols. A symbol is connected to its object because people agree on what it means (Fiske 2011, 44). It does not either reveal an implication of an idea or a physical resembles to an object. Its meaning is culturally accepted which takes a long time like numbers, the sign of heart meaning love, or signs of zodiac (Aiello 2020).

Thus, a player's choice of symbol in the game means that the player transfers his idea of a painting referring to its object, an IR theory statement, like a letter of an alphabet. The painting becomes a symbol if players agree on its representation of the concept. Thus, symbols emerge over time. A Rothko or a Pollock painting can respectively become symbols of Constructivism and Structural Realism after decades. The acceptance of zodiac signs, heart sign as symbols have emerged at the end of a long process that can be studied by evolutionary games.

6.5   Equilibrium

If players agree that a Rothko painting exactly corresponds to a theoretical statement and therefore the painting is an icon of the statement, then none of them would deviate from icon to index or to symbol given that the other insists that the painting constitutes an icon for that statement. The equilibrium of iconness constitutes a world of stability no player would deviate from under the constraint of interdependent consciousnesses.

An equilibrium indicates mental creations of semiotic relations between paintings and IR theories illustrating how mental states, that is, how different consciousnesses, beliefs, and knowledge interact (Chalmers 1997). Hence, VSGs divulge into communicative intricacies. The iconicity of a Rothko painting as an equilibrium for structural constraints corresponds to a constructed world (Goodman, 1978). It is a truth stemming from players' interdependent minds describing a myth "as a story by which a culture explains or understands some aspect of reality or culture" (Fiske 2011, 82). Players create a truth within the culture of the divided Discipline in turmoil. Hence, players transform the Discipline to some extent by achieving a coordination and creation of a common semiotic language.

Three Nash equilibria obtain in the generic Peircian game given in Fig. 6.1. Equilibrium strategy profiles fulfill the following conditions:

1. *(Icon, Icon)* pair of strategies constitutes a Nash equilibrium if and only if $a, b \geq 0$;
2. *(Index, Index)* pair of strategies constitutes a Nash equilibrium if and only if $c, d \geq 0$;
3. *(Symbol, Symbol)* pair of strategies constitutes a Nash equilibrium if and only if $e, f \geq 0$

The equilibria reflect incentives of players who merely want to coordinate their sign choices. Players both know that they will be rewarded for coordination of sign choices and each knows that the other is also trying to coordinate. Nevertheless, a coordination equilibrium can represent equal or unequal values for each alternative coordination equilibria can represent different values corresponding to players' varying happiness and satisfaction in each equilibrium. The equilibria allocate different payoffs to players according to possible preference orderings over the signs. The equilibrium remains constant but players' utilities differ in each case (Fig. 6.2):

If sign coordination ends up in equal sign payoffs for the players, then the question becomes the detection, disclosure of which sign has the conventional priority (Schelling 1983; Lewis 1969). If an equilibrium represents an indifference of players with respect to the type of the sign, that is, it does not matter whether they agree that a painting is an icon, index, or a symbol for a theoretical proposition so that signs' types are irrelevant and players obtain equal satisfaction in each equilibrium, then one can ask whether players prioritize one of them, say, symbol, over others.

For example, cases 4, 5, and 6 display that both Row and Column agree that the painting in question, be it one by Pollock or Rothko, constitutes a pictorial, graphical representation of structural constraints and dynamic nonunique anarchies, and, they prefer the sign of icon to index and symbol. The outcomes of (symbol, icon) and (icon, symbol) generate zero satisfaction due to the failure of players' coordination of sign choices. The (symbol, symbol) pair corresponds to the Column's view that the painting is a symbol, not an icon. The Row knows that her deviation to icon against the Column's claim implies a joint dissatisfaction in the game. Thus, the Row knows that if she chooses sign that establishes a visual correspondence between the painting and Wendt's claim while the Column insists that the painting constitutes a symbol, then they will both lose.

1	$a = c = e; b = d = f$	Players prefer the three signs equally.
2	$a = c > e; b = d > f$	Players prefer icon and index equally, but they prefer symbol the least.
3	$a = e > c; b = f > d$	Players prefer icon and symbol equally, but they prefer index the least.
4	$a > c = e; b > d = f$	Players prefer icon the most, they prefer index and symbol equally.
5	$a > c > e; b > d > f$	Players prefer icon to index, and they prefer index to symbol.
6	$a > e > c; b > f > d$	Players prefer icon to symbol, and they prefer symbol to index.
7	$c = e > a; d = f > b$	Players prefer index and symbol equally, but they prefer icon the least.
8	$c > a = e; d > b = f$	Players prefer index the most, they prefer icon and symbol equally.
9	$c > a > e; d > b > f$	Players prefer index to icon, and they prefer icon to symbol.
10	$c > e > a; d > f > b$	Players prefer index to symbol, and they prefer symbol to icon.
11	$e > a = c; f > b = d$	Players prefer symbol the most, they prefer icon and index equally.
12	$e > a > c; f > b > d$	Players prefer symbol to icon, and they prefer icon to index.
13	$e > c > a; f > d > b$	Players prefer symbol to index, and they prefer index to icon.

Fig. 6.2 Ranking of equilibrium payoffs in the generic Peircian game

The game can model two IR theorists' coordination on Pollock to constitute an icon, index, or a symbol for Wendt's anarchy conceptualization according to the payoff ordering of c > b > a. The most beneficial outcome is their agreement on symbol followed by icon and index. If either sign pair is reached, there is no reason for either player to deviate from it. If, for example, the Column chooses symbol, the Row's choice of icon or index leads to the failure of the coordination and consequently zero utilities for both.

6.6 Mixed-Strategy Equilibrium

Peirce (1965, 247) affirms that all signs are partly iconic (they denote by resembling their objects), indexical (they are "really affected" by their objects), and symbolic (they denote "by virtue of a law"). He admits a relation between icon, index, and symbol incorporates an index and an icon, and every index involves an icon. Thus, it is possible to support the view that a sign can be composed of various types permitting players to use mixtures of icon, index, and symbol in their interactions. For example, a painting can be an icon, index, and a symbol simultaneously, like 1/3 icon, 1/3 index, and 1/3 symbol. An example is the crossroad sign. The triangle sign is a symbol of warning, the cross is iconic for the crossroad, it is also an index because it indicates that one approaches a crossroads (Fiske 2011, 11, 46). Hence, a symbol can also mean an icon and an index. Icon, index, and symbol are not necessarily separated from each other in a clear-cut fashion. Consequently, Peirce's claim corresponds to a precise probability distribution over the pure strategies of symbol, index, and icon, therefore a mixed strategy.

To illustrate, suppose that a Rothko color-pane painting is selected as a cover of a book discussing Waltz's Structural Realism. Is the book cover an icon or a symbol? It is a symbol if the selection of the painting emanates from long discussions of the theory and some arguments advancing repeatedly that the painting constitutes a pictorial representation of Waltz's figure. The painting physically resembling to Waltz's figure becomes a symbol ultimately. Similarly, a player can think that a Pollock painting is partly an icon and partly an index for Wendt's anarchy arguments. These signs are fuzzy. There is a precise concept of strategy which corresponds to fuzzy signs in game theory: mixed strategies. Players can play strategies in Saussurean and Peircian games probabilistically corresponding to mixed strategies which can produce equilibria.

In the generic Peircian game given in the Fig. 6.1, let p_1, p_2, and $1 - p_1 - p_2$ denote Row's likelihoods of selecting icon, index, and symbol, and, similarly, let q_1, q_2, and $1 - q_1 - q_2$ denote Row's likelihoods of selecting icon, index, and symbol respectively. Players' indifference between signs is given by the following conditions:

Row is indifferent between icon, index and symbol if $q_1 a = q_2 c = (1 - q_1 - q_2) e$, solving for q_1, q_2, and $1 - q_1 - q_2$, we obtain $q_1 = c\,e/(c\,e + a\,c + a\,e)$, $q_2 = a\,e/(a\,e + a\,c + e\,c)$, and $1 - q_1 - q_2 = a\,c/(a\,c + e\,c + a\,e)$. Similarly, Column is indifferent between icon, index, and symbol if $p_1 b = p_2 d = (1 - p_1 - p_2) f$, solving for p_1, p_2, and $1 - p_1 - p_2$, we obtain $p_1 = d\,f/(d\,f + b\,d + b\,f)$, $p_2 = b\,f/(b\,f + d\,f + b\,d)$, and $1 - p_1 - p_2 = b\,d/(b\,d + d\,f + b\,f)$.

The mixed strategy can be interpreted as players' divisions about accepting the painting by Rothko or Pollock as an icon, index, or symbol for a proposition by Waltz or Wendt into three distinct groups. Thus, coordination between players takes three distinct directions. Each probability reflects players' beliefs about the painting constituting an icon, index, or symbol. For example, $q_1 = c\,e/(c\,e + a\,c + a\,e)$, $q_2 = a\,e/(a\,e + a\,c + e\,c)$, and $1 - q_1 - q_2 = a\,c/(a\,c + e\,c + a\,e)$ and $p_1 = d\,f/(d\,f + b\,d + b\,f)$, $p_2 = b\,f/(b\,f + d\,f + b\,d)$, and $1 - p_1 - p_2 = b\,d/(b\,d + d\,f + b\,f)$ are Column's and Row's beliefs about the painting being an icon, index, or symbol and are Row's beliefs, respectively. Thirdly, the mixed strategy can be interpreted as each pair of (p_1, q_1), (p_2, q_2), $(1 - p_1 - p_2, 1 - q_1 - q_2)$ constituting players' common knowledge expectations of the painting constituting icon, index, and symbol, respectively. If $a = 2$, $c = 1$, $e = 1$ and $b = 2$, $d = 1$, and $f = 2$, then these expectations for Row become the following: $p_1 = \frac{1}{4}$, $p_2 = \frac{1}{2}$, $1 - p_1 - p_2 = \frac{1}{4}$. Each value indicates Row's common knowledge expectations about icon, index, and symbol being signs. Similarly, Column's expectations become $q_1 = 1/5$, $q_2 = 2/5$, $1 - q_1 - q_2 = 2/5$. Variations in coordination payoffs yield variations in players' subjective expectations about signs. As result, the distribution of the players into three distinct sign groups changes with the magnitude of payoffs icon, index, and symbol generate.

6.7   Peircian Game in Extensive Form

A Peircian game in extensive form is a little more complex as the figure below shows (Fig. 6.3):

Row moves first by selecting between icon (*I*), index (*X*), and symbol (*S*). Column moves second by choosing between *I*, *X*, and *S* after Row's choice of *I*, by choosing between *I'*, *X'*, and *S'* after Row's choice of *X*, and by choosing between *I"*, *X"*, and *S"* after Row's choice of *S*. If the outcome is (*I, I*), Row obtains a payoff of *a* Column obtains a payoff of *b*, if the outcome is (*X, X'*), Row obtains a payoff of *c* and Column obtains a payoff of *d*, and if the outcome is (S, S"), Row obtains a payoff of *e* Column obtains a payoff of *f*. It is assumed that *a, b, c, d, e, f > 0*. Players obtain zero payoffs if they do not reciprocate mutual choices by selecting the same signs. Column has perfect information: she knows Row's previous sign choice before she selects hers. Column's responses are in the same form as I, X, and S. They are distinguished as her responses to Row's choice of icon, index, and symbol.

In the particular game given by the Fig. 6.3, Row obtains the payoff of *a = 2* and Column obtains the payoff of *b = 1*, if both choose index, Row obtains the payoff of *c = 1* and Column obtains the payoff of *d = 2*, if both choose symbol, Row obtains the payoff of *e = 0* and Column obtains the payoff of *f = 2*. If they select different signs, Row and Column obtain zero payoffs.[2]

The game is solved by backward induction, therefore Column's reaction to Row's choices is established first at the end outcomes of the game. Column always reciprocates Row's choices, because she obtains no utility otherwise. Reciprocation dominates non-reciprocation at each of her three information sets. Hence, if Row chooses *I*, she responds by *I*, if Row chooses *X*, she responds by *X'*, and If Row chooses S, she responds by *S"*. Row selects between *I*, *X*, and *S* knowing that Column always reciprocates her action. Row then has the power of determining the

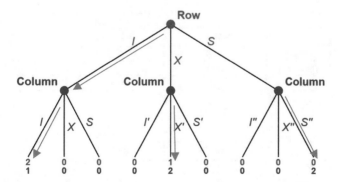

Fig. 6.3 Peircian game in extensive form

---

[2] Different payoffs to reciprocated outcomes generate different Peircian games keeping non-reciprocated payoff outcomes equal to zero, for example.

outcome of the interaction. She picks the action generating her best outcomes as long as she has strict preferences over the outcomes for signs and she is not indifferent between them. If Row is indifferent between two signs yielding equal payoffs, then the game is not solvable by backward induction. An example is, for example, Row's indifference among the three signs. The game is however solvable if Row's best utility lies in a specific sign, but she is indifferent between signs representing lesser but equal utilities. The game given in the figure above gives the following solution: given Column's reactions, Row prefers I that gives the maximum payoff of 2. The equilibrium becomes (I, I).

The Peircian game in extensive form teaches the lesson of the power of the player who moves first. Column has a weak position as she must agree with the Row's sign choice as she reciprocates. Column might, for example, prefer symbol the most as the sign of the painting but if Row prefers index the most, then she has to accept, Row's choice and concur on the painting's meaning being an index. This is a price the last moving player has to pay to achieve coordination in a Peircian game.

6.8   One-Sided Bayes-Nash Equilibrium

Suppose that Row believes Column is not interested in opening a new communication channel through art. We can express this idea by positing Row's beliefs about Column's preferences similar to Bayesian Saussurean games. Row's misperception is summarized in his belief that Column is indifferent towards sign making so that Column payoffs are equal to zero in all cells of the Peircian game with a probability of π and that Column is not indifferent toward sign making with probability of $1 - \pi$ as in the fig. 6.4:

Row has unique type; therefore, his strategy set is $\{I, X, S\}$. Column has two types: type 1 and type 2. Column's type contingent strategy set amounts to the

		Column		
		Icon (I)	Index (X)	Symbol (S)
Row	Icon (I)	a, 0	0, 0	0, 0
	Index (X)	0, 0	c, 0	0, 0
	Symbol (S)	0, 0	0, 0	e, 0

(π)

		Column		
		Icon (I)	Index (X)	Symbol (S)
Row	Icon (I)	a, b	0, 0	0, 0
	Index (X)	0, 0	c, d	0, 0
	Symbol (S)	0, 0	0, 0	e, f

$(1 - \pi)$

Fig. 6.4 Incomplete information Peircian game

	I	X	S
I, I	$a.\pi + (1-\pi).a = a$	$0.\pi + (1-\pi).0 = 0$	$0.\pi + (1-\pi).0 = 0$
I, X	$a.\pi + (1-\pi).0 = a\pi$	$0.\pi + (1-\pi).c = (1-\pi)c$	$0.\pi + (1-\pi).0 = 0$
I, S	$a.\pi + (1-\pi).0 = a\pi$	$0.\pi + (1-\pi).0 = 0$	$0.\pi + (1-\pi).e = (1-\pi)e$
X, I	$0.\pi + (1-\pi).a = (1-\pi)a$	$c.\pi + (1-\pi).0 = c\pi$	$0.\pi + (1-\pi).0 = 0$
X, X	$0.\pi + (1-\pi).0 = 0$	$c.\pi + (1-\pi).c = c$	$0.\pi + (1-\pi).0 = 0$
X, S	$0.\pi + (1-\pi).0 = 0$	$c.\pi + (1-\pi).0 = c\pi$	$0.\pi + (1-\pi).e = (1-\pi)e$
S, I	$0.\pi + (1-\pi).a = (1-\pi)a$	$0.\pi + (1-\pi).0 = 0$	$e.\pi + (1-\pi).0 = e\pi$
S, X	$0.\pi + (1-\pi).0 = 0$	$0.\pi + (1-\pi).c = (1-\pi)c$	$e.\pi + (1-\pi).0 = e\pi$
S, S	$0.\pi + (1-\pi).0 = 0$	$0.\pi + (1-\pi).0 = 0$	$e.\pi + (1-\pi).e = e$

Fig. 6.5 Row's expected payoff table

Cartesian product of $\{I, X, S\} \times \{I, X, S\} = \{(I, I), (I, X), (I, S), (X, I), (X, X), (X, S),$ $(S, I), (S, X), (S, S)\}$. The first letter in each parenthesis refers to the choice of type 1 and the second one to the one of type 2. The next step is to construct Row's expected payoff table. Column has no dominant action; thus, the table will contain three Columns corresponding to Row's strategies of I, X, and S and nine Rows corresponding to Column's type-contingent strategies generating 29 Bayes-Nash equilibrium candidates. The table is given below Fig. 6.5:

1. $\{I, (I, I)\}$: Row's choice of I against Column's type-contingent strategy of (I, I) generates a positive payoff of a unlike his strategies of X and S generating a payoff of 0. Hence, Row's choice of I is the best reply to (I, I) regardless Row's belief about the Column's type. Type 1 is indifferent between her choice of I and her choices of X and S. Thus, I is the best reply for type 1 against Row's choice of I. Type 2 receives a payoff of $b > 0$ from her choice of I but 0 from her choices of X and S. Therefore, type 2 either has no incentive to deviate from I to X or S facing Row's choice of I. We obtain the first Bayes-Nash equilibrium as result in the form of: $\{I, (I, I)$; for beliefs π of Row$\}$.

2. $\{I, (I, X)\}$: Row never chooses S against (I, X) because S generates a zero payoff unlike I and X against (I, X). Row's choice of I is the best reply against (I, X) if a $\pi \geq (1 - \pi) c$ implying the condition of $\pi \geq c/(c + a)$. Assume that this belief condition fulfilled. Thus, Row responds to (I, X) by selecting I. Type 1 has no incentive to shift from I to either X or S. As to type 2, she has an incentive to deviate from X to I because $b > 0$. Hence, no Bayes-Nash equilibrium can contain the strategy profile of $(I, (I, X))$. Row's choice of X is the best reply against

(I, X) if $a\,\pi \le (1 - \pi)\,c$ implying the condition of $\pi \le c/(c + a)$. Assume that this belief condition fulfilled. Thus, Row responds to *(I, X)* by selecting X. Type *1* has no incentive to shift from *I* to either X or S. As to type 2, she has no incentive to deviate from X to *I* or S because $d > 0$. The result is the second Bayes-Nash equilibrium in the form of $\{X, (I, X); \pi \le c/(c + a)\}$.

3. $\{I, (I, S)\}$: Row never chooses X against *(I, S)* because X generates a zero payoff unlike *I* and S against *(I, S)*. Row's choice of *I* is the best reply against *(I, S)* if $a\,\pi \ge (1 - \pi)\,e$ implying the condition of $\pi \ge e/(e + a)$. Assume that this belief condition fulfilled. Thus, Row responds to *(I, S)* by selecting *I*. Type *1* has no incentive to shift from *I* to either X or S. As to type 2, she has an incentive to deviate from S to *I* because $b > 0$. No Bayes-Nash equilibrium can contain the strategy profile of *(I, (I, S))*. Row's choice of *I* is the best reply against *(I, S)* if $a\,\pi \ge (1 - \pi)\,e$ implying the condition of $\pi \ge e/(e + a)$. Assume that this belief condition fulfilled. Type 1 is indifferent and thus has no incentive to deviate but type 2 does because not S but *I* is the best reply against Row's choice of *I*. Hence, no Bayes-Nash equilibrium obtains in this case either.

4. $\{I, (X, I)\}$: Row never chooses S against *(X, I)* because S generates a zero payoff unlike *I* and X against *(X, I)*. Row's choice of *I* is the best reply against *(X, I)* if $(1 - \pi)\,a \ge c\,\pi$ implying the condition of $\pi \le a/(a + c)$. Assume that this belief condition fulfilled. Thus, Row responds to *(X, I)* by selecting *I*. Type *1* has no incentive to shift from X to either *I* or S. As to type 2, she has no incentive to deviate from *I* to X either because $b > 0$. We obtain the third Bayes-Nash equilibrium in the form of: $\{I, (X, I); \pi \le a/(a + c)\}$ as result. Row's choice of X is the best reply against *(X, I)* if $(1 - \pi)\,a \le c\,\pi$ implying the condition of $\pi \ge a/(a + c)$. Assume that this belief condition fulfilled. Against Row's choice of X type 2 would deviate from *I* to X while type 1 is indifferent. Hence, no equilibrium in this case.

5. $\{I, (X, X)\}$: Row never chooses *I* or S against *(X, X)*. His best reply is X against *(X, X)* unconditional upon his beliefs about Column's type. Type *1* has no incentive to deviate from X due to her indifference. Type *2* either has no incentive to deviate from X to either *I* or S to get zero payoff while X generates a positive payoff of $d > 0$. The result is the fourth Bayes-Nash equilibrium in the game in the form of $\{X, (X, X); \text{for beliefs } \pi \text{ of Row}\}$.

6. $\{I, (X, S)\}$: Row never chooses *I* against *(X, S)*. His best reply to *(X, S)* is X if $c\,\pi \ge (1 - \pi)\,e$ implying the belief condition of $\pi \ge e/(e + c)$. Suppose that this condition is satisfied so that Row selects X. Type 1 has no incentive to deviate from X either to *I* or S. Type 2 however would deviate from S to X. Therefore, no Bayes-Nash equilibrium exists in this case. Now, suppose that $\pi \le e/(e + c)$, thus Row selects S. Type 1 is indifferent, so that she has no incentive for any deviation but type 2 would respond by S against Row's choice of S because her best reply against Row's choice of S is S. The result is the fifth Bayes-Nash equilibrium in the game in the form of $\{S, (X, S); \pi \le e/(e + c)\}$.

7. $\{I, (S, I)\}$: Row never chooses X against *(S, I)*. He selects *I* against *(S, I)* if $(1 - \pi)\,a \ge e\,\pi$ implying the condition of $\pi \le a/(a + e)$. Assume that this belief condition fulfilled. Row selects *I*. Type 1 has no incentive to deviate from S and type 2

either has no incentive to deviate from I given that Row chooses I. The result is the sixth Bayes-Nash equilibrium in the form of $\{I, (S, I); \pi \le a/(a + e)\}$. Row selects S against (S, I) if $(1 - \pi) a \le e \pi$ implying the condition of $\pi \ge a/(a + e)$. Suppose that this condition holds. Row selects S against (S, I). While type 1 is indifferent so that she has no deviation incentive, type 2 would shift from I to S. Therefore, no Bayes-Nash equilibrium exists containing the profile $(S, (S, I))$.

8. $\{I, (S, X)\}$: Row never chooses I against (S, X). He selects X against (S, X) provided that $(1 - \pi) c \ge e \pi$ implying the condition of $\pi \le c/(c + e)$. Assume that belief condition is satisfied. Row chooses X against (S, X). Type 1 has no incentive to deviate from S. Type 2 either has no incentive to deviate from X to coordinate with Row's choice of X. Therefore, the seventh Bayes-Nash equilibrium is obtained under these conditions in the form of $\{(X, (S, X); \pi \le c/(c + e)\}$. Row selects S against (S, X) provided that $(1 - \pi) c \le e \pi$ implying the condition of $\pi \ge c/(c + e)$. Type 1 does not deviate from S, but type 2 deviates from X to S. Thus, no Bayes-Nash equilibrium obtains in the form of $\{(X, (S, X); \pi \ge c/(c + e)\}$.

9. $\{I, (S, S)\}$: Row never chooses either I or X against Column's strategy of (S, S) but S regardless his beliefs. Type 1 is indifferent; hence she would not deviate from S. Type 2 has no incentive to deviate from S to either I or X. As result the eighth Bayes-Nash equilibrium is obtained in the form of $\{S, (S, S); $ for all beliefs of $\pi\}$.

The Bayes-Nash equilibria obtained imply that in three instances Row's misperception of Column's type does not matter provided that Row and both types select identical actions of icon, index, or symbol. Row's beliefs start to matter only if Column's types select different actions. Start with the second equilibrium. The belief threshold of $c/(c + a)$ informs that as Row's payoff from a coordination on the iconicity of a painting increases, Row's belief that he interacts with the indifferent Column type must get smaller and therefore Row's belief that that he interacts with the Column interested in coordination must get higher for the equilibrium on the indexity of the painting to form.

In the third equilibrium, the iconicity becomes the stable outcome only if Row's belief that he interacts with the indifferent type remains below the threshold of $a/(a + c)$. The belief condition implies the threshold gets smaller as Row's payoff of indexity of the painting increases. Therefore, Row is inclined to select icon provided that he believes he is interacting with type 2 rather than with type 1 who is not interested in generating theoretical meanings of the sentence either by Waltz or Wendt through Rothko or Pollock artworks. The higher Row's payoff to indexity becomes, the higher becomes the propensity of Row to select icon in the game.

In the fifth equilibrium which depends on a condition of Row's belief about Column's type, the painting in question becomes accepted as a symbol if Row perceives that he interacts with type 2 provided that the indexity of the painting gains higher values as the threshold $e/(e + c)$ diminishes due to higher values of c.

The sixth equilibrium implies that as Row's payoff to the symbolicity e increases, smaller beliefs of Row of interacting with the indifferent type prompt Row to reciprocate the icon choice of type 2. Hence, the value of the painting as a symbol plays a role in Row's selection of the painting as an icon. Finally, the seventh equilibrium implies that increments in Row's payoff of symbolicity e allows for higher likelihoods of Row's reciprocation of type 2's index choice.

The equilibria depending on Row's beliefs teach also that Row's payoff parameter in both the nominator and the denominator of respective belief conditions is equally effective in fulfilling equilibrium conditions. Row's payoffs to a coordination in iconicity, indexity, and symbolicity generate higher thresholds and therefore smaller ranges for Row's belief of interacting with type 2 to reach equilibria.

6.9   Evolutionary Equilibrium

Peirce notes that symbols come up only after a long process. Therefore, the concept of evolutionarily stable equilibrium (ESS) in VSGs must be introduced as well. The latter equilibrium concept is applied in evolutionary games and does not require existence of rational players. The evolutionarily stable equilibrium (ESS) exposes stability conditions in populations of IR scholars in Saussurean and Peircian games like whether the whole population accept, for example, that a Rothko painting constitutes a signifier or an icon for Waltz's structural constraints; otherwise, a "mutant" mind can prompt a dynamic process of conversion of the whole population to accept an alternative meaning. The third part is Chap. 5 that interprets results obtained in terms of different schools of philosophy of mind. The central questions revolve around how players' mental states and consciousnesses get aligned with the rules of VSGs.

Evolutionary Peircian games constitute an alternative avenue to develop VSGs to study the emergence of paintings not only as symbols but as icons or indices as well in Peircian games. The strategies of icon (I), index (X), and symbol (S) are interpreted as modes of behavior, that is, modes of sign making evolving through a process of selection. The adoption of one type of sign depends on changing proportions of two other signs adopted in the population. For example, icon becomes an ESS (evolutionarily stable strategy) depending on how successful are alternative signs of index and symbol in the population. Suppose that icon is an ESS. Icon is an ESS if and only if a small proportion of the population adopt index or symbol, then the selection process will eliminate index and symbol ultimately. Icon, as an ESS, resists such mutations. An ESS is a Nash equilibrium satisfying the additional property of resistance to mutant invasions. John Nash has interpreted his equilibrium concept as "mass action," that is, interactions in populations in statistical terms (Björnerstedt and Weibull 1994). Players are repeatedly and randomly matched in large populations. They are not required to know that they are involved in a game; they are not able to solve optimization problems in their interactions.

The ESS in the Peircian game corresponds to Nash equilibria as Nash equilibria are strict. Recalling the definition of evolutionarily stable strategy by Maynard Smith (1982, 14) the implications are the following in a Peircian game:

I is an ESS if:

1. $E(I, I) > E(X, I)$ and
2. $E(I, I) > E(S, I)$

X is an ESS if:

3. $E(X, X) > E(I, X)$ and
4. $E(X, X) > E(S, X)$

S is an ESS if:

5. $E(S, S) > E(I, S)$ *and*
6. $E(S, S) > E(X, S)$

All of *conditions* above are satisfied. Therefore, all three signs are ESS. Population dynamics, depending on payoffs, drive to the three stability sink points consisting of *I*, *X*, and *S*.

Suppose that the Peircian game is (Fig. 6.6):

The phase diagram of the Peircian game produced by the Python code for evolutionary games is produced below (Figs. 6.7 and 6.8):

		Column		
		Icon (I)	*Index (X)*	*Symbol (S)*
Row	*Icon (I)*	*1, 2*	*0, 0*	*0, 0*
	Index (X)	*0, 0*	*2, 1*	*0, 0*
	Symbol (S)	*0, 0*	*0, 0*	*2, 2*

Fig. 6.6   A Peircian evolutionary game

Fig. 6.7   Phase Diagram 1
of the Peircian
evolutionary game

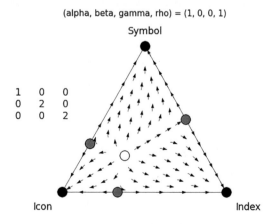

(alpha, beta, gamma, rho) = (1, 0, 0, 1)

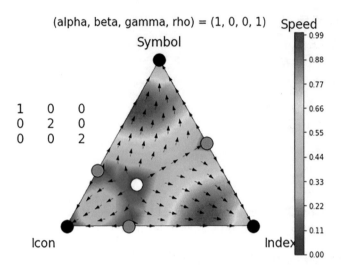

Fig. 6.8 Phase Diagram 2 of the Peircian evolutionary game. (I am grateful to Selcen Pehlivan for her help in writing the Python code for evolutionary games)

The phase diagram shows that the attraction basin of the ESS of (*I, I*) is the smallest compared to those (*X, X*) and (*S, S*) ESSs. The evolutionary stability demonstrates that any population adopting a mixture of I, X, and S is unstable. No polymorphic equilibrium therefore exists. It is impossible that the population will finally reach stability with some adopting *I*, some adopting *X* and some adopting *S*. Only one of the signs survives interactions while the other two die out depending on which attraction basins they are located. Changing the payoff structure of the game there arise alternative ESSs and attraction basins. Evolutionary VSGs are useful to ponder about conditions under which symbol becomes a stable norm in a population of IR theorists with symbol being assumed to take several generations to be accepted. They teach that not only symbols, but signs of icon and index as well might take several generations to be established as stable conventions for interpretations of paintings and IR theories forming signs together.

Chapter 7
Hybrid Games

One of the basic criticisms targeting game theory is that players' preferences do not change over the course of the interaction. Wendt (1992, 416) notes that "In the traditional game-theoretic analysis of cooperation, even an iterated one, the structure of the game – of identities and interests – is exogenous to the interaction and, as such, does not change." Wendt's remark is about finite or infinite interactions where the same game is repeated. If different games follow one another, then it is possible to model changes in preferences.

I propose a two-stage game with a changing structure. It allows players to interact in a Saussurean game in the first round and in a Peircean game in the second. Players first interact to decide whether a painting constitutes a signifier. If they concur that the painting is not a signifier, then the game ends and players receive payoffs of their joint disagreement. The end of the game is interpreted either as players prefer an alternative painting or they oppose the idea of developing a common semiotic language to evaluate statements by Waltz and Wendt. Otherwise, if they accept, then they reach the second stage consisting of a Peircian game. They decide on the nature of the sign in the second stage, that is, they decide whether the painting and the proposition form either an icon or an index or a symbol. The second-stage game ends once players coordinate their strategies accepting that the painting constitutes one of the three signs. Hence, players know that their rejection of the painting as a signifier means the missed opportunity of a semiotic language and that their acceptance does not allow any rejection of the Saussurean game models. Once the second stage of the game is reached there is no way out; players must agree on one of the three signs. The game is qualified as being hybrid because it connects Saussurean and Peircian games.

© Springer Nature Switzerland AG 2023
S. Ş. Güner, *Art and IR Theory*, Mathematics in Mind,
https://doi.org/10.1007/978-3-031-32342-3_7

7.1 Hybrid Game Under Imperfect Information

Would the equilibria change if players have imperfect information? This question is answered by the hybrid game in strategic form. Players' strategy spaces depend on their actions at Saussurean and Peircian games. Row's strategy set is *{A, R}* × *{I, X, S}* = *{AI, AX, AS, RI, RX, RS}*. Column's strategy set is *{A, R}* × *{A', R'}* × *{I, X, S}* × *{I'X', S'}* × *{I", X", S"}* = *108*. Thus, when translated into the strategic form, the hybrid game of perfect information becomes a 6 × 108 matrix generating multiple Nash equilibria open to reduction and refinement. Imperfect information assumption renders the hybrid game more manageable as strategic form abstracts information conditions away. Players move ignoring previous choices other players have made in the game. Hence, the hybrid game of imperfect information rules are as follows: Row moves first by selecting either *A* or *R* followed by Column who, ignoring Row's choice, also chooses either *A* or *R*. The game reaches the Peircian game stage only if both players choose A in the Saussurean game. In the Peircian game, Row then selects either *I*, or *X*, or *S*. Column chooses either *I*, or *X*, or *S* ignoring what exact choice Row's has made. The game ends by Column's choice. To recall, *(R, R')* outcome generates a payoff of *α* for Row and *β* for Column. The outcome *(I, I)* gives a payoff of *a* and a payoff of *b* to Column, *(X, X)* gives a payoff of *c* to Row and *d* to Column, and the outcome of *(X, X)* yields a payoff to *e* to Row and *f* to Column. All outcomes resulting of miscoordination generate *0* payoff for both players.

Row's strategy set is the Cartesian product of her actions at her two information sets: *{A, R}* × *{I, X, S}* = *{AI, AX, AS, RI, RX, RS}*. Column's strategy set is the Cartesian product of her choices in the Saussurean and the Peircian stages, namely: *{A, R}* × *{I, X, S}* = *{AI, AX, AS, RI, RX, RS}*. The hybrid game of imperfect information is a 6 × 6 strategy matrix as shown below (Fig. 7.1).

Players choose strategies ignoring strategy choices of each other. The game exposes 12 Nash equilibria out of which nine of them produce the *(R, R)* outcome, namely, *{(RI, RI), (RI, RX), (RI, RS), (RX, RI), (RX, RX), (RX, RS), (RS, RI), (RS, RX), (RS, RS)}*. There exist an additional three other equilibria demonstrating players' coordination on icon, index, and symbol, namely *{(AI, AI), (AX, AX), (AS, AS)}*. Hence, the game implies that either players agree to not to coordinate in nine diverse ways or agree to coordinate in three diverse ways.

The multitude of equilibria implies weaker explanations. The lesser the number of equilibria, the more powerful become explanations. There is a way to reduce the equilibria by the refinement of Nash equilibria by the concept of subgame-perfect Nash equilibrium. A Nash equilibrium becomes subgame perfect if it generates Nash equilibrium in every subgame. The hybrid game has two subgames: the whole game and the Peircian game. The Saussurean game is not a subgame, because it progresses to the Peircian game if both players select *A*. But the Peircian game is subgame, because Row's information set is a singleton, and all successor nodes are connected to the singleton information set of Row. Saussurean game does not end if both players select *A* and in that case the hybrid game reached the Peircian second

		Column					
		AI	AX	AS	RI	RX	RS
	AI	a, b	0, 0	0,0	0,0	0, 0	0,0
	AX	0, 0	c, d	0, 0	0, 0	0, 0	0, 0
Row	AS	0, 0	0, 0	e, f	0, 0	0, 0	0, 0
	RI	0, 0	0, 0	0, 0	α, β	α, β	α, β
	RX	0, 0	0, 0	0, 0	α, β	α, β	α, β
	RS	0, 0	0, 0	0, 0	α, β	α, β	α, β

Fig. 7.1 Hybrid game in imperfect information

		Column	
		Accept (A)	Reject (R)
	Accept (A)	Next stage	0, 0
Row	Reject (R)	0, 0	α, β

Fig. 7.2 Hybrid game period 1

state. Thus, the Saussurean game is not a game that can be analyzed as if it is a game on its own. Accordingly, all Nash equilibria ending in disagreement in the Saussurean game are eliminated; none of the equilibria in the set {*(RI, RI), (RI, RX), (RI, RS), (RX, RI), (RX, RX), (RX, RS), (RS, RI), RS, RX, RS, RS}* constitutes a subgame-perfect equilibrium. The result means that even if both players prefer to disagree because disagreement payoffs of α and β are higher than agreement payoffs of a, b, c, d, e, and f, strategy profiles generating disagreement are not subgame perfect. Hence, the subgame perfect Nash equilibrium that reduces the number of equilibria from 12 to 3. Players coordinate on selection of icon, index, or symbol.

7.2   Hybrid Game in Two-Stage Form

It is possible to present the hybrid game as a two-stage game as below (Gibbons 1992, 74–75) (Figs. 7.2 and 7.3):

		Column		
		Icon	*Index*	*Symbol*
Row	*Icon*	*a, b*	*0, 0*	*0, 0*
	Index	*0, 0*	*c, d*	*0, 0*
	Symbol	*0, 0*	*0, 0*	*e, f*

Fig. 7.3 Hybrid game period 2

		Column	
		Accept (A)	*Reject (R)*
Row	*Accept (A)*	*a, b*	*0, 0*
	Reject (R)	*0, 0*	*α, β*

Fig. 7.4 Hybrid game final period

We substitute the equilibrium reached in the second period into the game matrix at the first period. If $a > c \geq e$, for example, that is, Row prefers icon to index and symbol, the game matrix becomes (Fig. 7.4):

The game has two equilibria occurring in the *(A, A)* and *(R, R)* profiles. The *(R, R)* equilibrium means that both players reject the painting as a signifier of a theoretical proposition at the first stage. If the game advances to the second stage, then the equilibrium implies both players' acceptance of the painting as an icon. Column is forced to agree with Row because her best reply is icon against Row's choice of icon in the Peircian game. In fact, Column's choices run in the direction of reciprocation of Row's preferences over icon, index, and symbol. For example, Column would prefer that the sign becomes index or symbol but never icon. Yet, as Row selects her dominant strategy of icon following the payoff assumption of $a > c > e$, in the Peircian game it is not optimal for Column to insist on index or symbol.

7.3   Hybrid Game of One-Sided Incomplete Information

The hybrid game in the final period is useful to investigate implications of asymmetric information among players. Suppose that Column misperceives Row; she is unsure whether Row prefers icon or symbol in the final Peircian stage of the hybrid game. Hence, the equilibrium reached in the second stage consisting of the Peircian game is nested in the Saussurean game of the first stage. Similar to the previous analysis of one-sided incomplete information in a Saussurean game, suppose that Column has a unique type and misperceives Row as preferring either icon or symbol with probabilities of p and $1 - p$. Column believes that coordination realizes at the equilibrium of *(I, I)* or the equilibrium of *(S, S)*. Thus, Row has two types: Type 1 prefers icon, Type 2 prefers symbol. The Column's updated misperception

		Column	
		A	R
	A	a, b	0, 0
Row	R	0, 0	a, β

$$[\pi]$$

		Column	
		A	R
	A	e, f	0, 1
Row	R	0, 0	a, β

$$[1 - \pi]$$

Fig. 7.5 Bayesian hybrid game

probabilities become π and $1 - \pi$, that is, Column believes that she is interacting with Type 1 with a likelihood of π and with Type 2 with the likelihood of $1 - \pi$. Let utilities be equal to the extensive form game of perfect information. The Bayesian game based on the final period of the hybrid game is given below (Fig. 7.5):

Players' strategies take different meanings in the two-stage hybrid game. Column has two strategies due to her unique type: A and R; she accepts or rejects the sign. The strategy of A means that "reciprocate Row's sign choice" and R means that "reject Row's sign choice." Row, misperceived by Column, has four type-contingent strategies of (A, A), (A, R), (R, A), (R, R). Row's type-contingent strategy of (A, A) corresponds to "I agree on the sign regardless my type," (A, R) means that "agree on the sign if I am Type *1* but reject the sign if I am Type 2, (R, A) means that "I reject the sign if I am Type *1* and accept the sign if I am Type 2," (R, R) means that "I reject the sign regardless my type." We have to first compute Column's best replies under her incomplete information and find out whether any of the two types of Row has any incentive to deviate from her preferred actions. If all types and Column has no incentive to deviate from her strategies, then we arrive at a Bayes-Nash equilibrium in the hybrid game.

Column's expected payoff matrix displaying her payoffs to her strategies A and R as pitted against Row's type-contingent strategies (A, A), (A, R), (R, A), (R, R) is (Fig. 7.6):

7.3.1   (A, AA)

If $\pi b + (1 - \pi) f \geq 0$, then Column's choice of A constitutes the best reply to Row's strategy of (A, A). The inequality implies a threshold condition upon Column's belief that Row prefers icon over symbol: $f/(f - b) \geq \pi$. If Column values icon more than symbol, so that $b > f$, then $f/(f - b)$ is negative. A negative value cannot be higher than or equal to a probability; the condition cannot be fulfilled as $0 \leq \pi \leq 1$. Therefore, A cannot constitute the best reply against (A, A). No Bayes-Nash

	A	R
AA	$\pi b + (1 - \pi) f$	0
AR	πb	$(1 - \pi) \beta$
RA	$(1 - \pi) f$	$\pi \beta$
RR	0	β

Fig. 7.6 Column's expected payoff matrix Column's expected payoff matrix

equilibrium is possible. Otherwise, if $b < f$, that is, if Column prefers coordination on symbol, then the threshold $f/(f - b)$ is positive and exceeds 1. All probabilities are smaller than 1 or equal to 1 at their maximum value. Therefore, the condition $f/(f - b) \geq \pi$ can be satisfied. Column responds by selecting A against (A, A), that is, both types selecting A. Would any of the two types have any incentive to deviate from A to R facing Column's choice of A? The answer is no, because both types would then obtain a payoff of zero instead of a and e that are strictly positive. Therefore, a second equilibrium under two conditions arises: Column prefers coordination over symbol to coordination over icon, and her belief of interacting with the Type 1 is sufficiently small: $\{A, (AA); f/(f - b) \geq \pi\}$.

7.3.2 (A, AR)

For A to be the best reply against (A, R) we must have $\pi b \geq (1 - \pi) \beta$ implying the following condition upon Column's belief $\pi \geq \beta/(b + \beta)$. Suppose the condition holds. Would any type of Row have an incentive to deviate against Column's choice of A? The answer is yes; Type 2 would shift from R to A against Column's choice of A. No Bayes-Nash equilibrium exists incorporating the profile (A, AR).

7.3.3 (A, RA)

If $(1 - \pi) f \geq \pi \beta$, that is, if $\pi \leq f/(f + b)$, then A is the best reply against (R, A) for Column. Suppose that the condition $\pi \leq f/(f + b)$ holds. Therefore, Column selects A against (R, A). Type 1 would deviate from R to A facing Column's choice of A. No equilibrium results in this case.

7.3.4   (A, RR)

Against (R, R), not A but R constitutes the best reply for Column because $\beta > 0$. Therefore, no equilibrium exists in this case.

7.3.5   (R, AA)

Against (A, A), R is the best reply if $0 \geq \pi\, b + (1 - \pi)\, f$. The inequality implies the condition of $\pi\,(f - b) \geq f$. If $f \leq b$, then it must be the case that $0 \geq f$. However, $f > 0$ by assumption. Otherwise, if $f > b$, then $\pi\,(f - b) \geq f$ cannot be satisfied because $\pi \geq f/(f - b)$ implies that π is higher than a threshold that is higher than 1. No belief condition can satisfy Bayes-Nash equilibrium for this strategy profile. Therefore, there is no Bayes-Nash equilibrium.

7.3.6   (R, AR)

If $\pi\, b \leq (1 - \pi)\, \beta$, that is, if $\pi \leq \beta/(\beta + b)$, then R is the best reply against (A, R). Suppose that $\pi \leq \beta/(\beta + b)$ holds. Column chooses R against (A, R). However, Type 1 would deviate from R to A given that Column chooses R. Therefore, no equilibrium is reached under these conditions.

7.3.7   (R, RA)

If $(1 - \pi)\, f \leq \pi\, \beta$, that is, if $\pi \geq f/(f + \beta)$, then R is the best reply against Row's type-contingent strategy of (R, A) for Column. Suppose that the condition $\pi \geq f/(f + \beta)$ holds. Therefore, Column selects R against (R, A). While Type 1 has no incentive to deviate from R against Column's R choice, Type 2 would shift from A to R facing Column's choice of R to avoid the payoff of zero and obtain the payoff of α. No equilibrium results including this strategy profile.

7.3.8   (R, RR)

The equilibrium candidate (R, RR) is the only profile that does not depend on Column's updated beliefs. It is assumed that $\beta > 0$, therefore R is the best reply against (R, R). None of the Row types has an incentive to deviate from R against R as $\alpha > 0$. Therefore, $\{R, (R, R); \text{for all beliefs of Column}\}$ constitutes a Bayes-Nash equilibrium independent of beliefs of π of Column. Coordination on rejection is achieved.

In summary, if Column misperceives Row as having different preferences over the types of signs, Column's misperceptions do not enrich equilibrium analysis further; two Bayes-Nash equilibria do not depend on Column's misperceptions. As Row is in the symmetric situation, the same holds for Row. And if two-sided incomplete information is assumed, again the result of beliefs having no impact upon equilibrium strategies is expected to occur. Yet, there is a point the Bayes-Nash equilibrium concept reveals: Column must prefer coordination over symbol to coordination over icon for both Row types and Column accept symbol as the sign. Without making any preference assumption concerning icon over symbol, it is impossible to compute an equilibrium. The equilibrium requires an extra information about Column's preferences over icon and symbol. The hybrid game of one-sided incomplete information implies two equilibria both not depending upon beliefs. One equilibrium forces the model to answer the question of "Does Column prefer symbol over icon?" If the answer is affirmative, then the game offers a different insight about what the strategic interdependence between a misperceiver and a misperceived player implies for their best replies in sign making. The insight corresponds to the joy of offering a model and the model's asking the modeler to make an additional supposition. What joy would a modeler feel? Is it mental so that it emanates from her mind that expands in space or from something else? This type of question will be the subject matter of the next chapter.

7.4 Hybrid Game in Extensive Form

The hybrid game in extensive form given below is solved by backward induction (Fig. 7.7).

Column who moves the last in the Peircian game reciprocates sign choices of Row. Branches connecting different sign choices are pruned off the tree. Row prefers to choose icon (I) that generates her highest payoff of 2 compared to index (X) generating a payoff of 0 and symbol (S) generating a payoff of 1 given Column's sign choices. Row choices of X and S are pruned off the tree. Column who decides whether the game develops into the Peircian stage or ends at Saussurean stage chooses A that gives her a payoff of 1 of the icon equilibrium compared to her payoff of 0 resulting from the coordination failure in the Saussurean stage. Hence, Column's R choice pruned off the tree. Back at the initial node of the game, Row who foresees all these sequences of choices, reasons back, and selects A to obtain 2. Consequently, the sequence of equilibrium actions becomes the following: Row chooses A at the start, Column responds by A, Row chooses I, and Column chooses I. The equilibrium sign becomes icon.

The game above is a special case. Changed payoffs lead to different hybrid games. In general, it is assumed that if Column reacts to Row's choice of R by R,

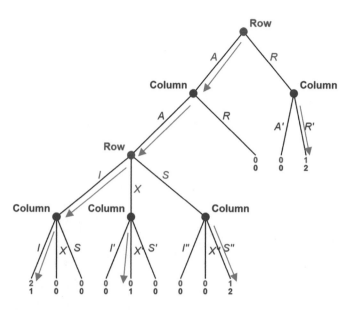

Fig. 7.7 Hybrid game in extensive form

then coordination in rejection is achieved creating payoffs of $\alpha > 0$ for Row and $\beta > 0$ for Column. If Column responds to Row's choice of A by selecting R, both players obtain zero and game ends. If α is strictly greater than any coordination payoff for Row in the Peircian game, then Row selects R. The equilibrium then becomes the pair (R, R). If α is strictly smaller than any coordination payoff for Row in the Peircian game, then Row selects A forcing Column to reciprocate and permitting the game to evolve into the Peircian stage. Row, following Column's reciprocation of her choice of A, moves by choosing between icon, index, and symbol. Row selects the sign Column selects under the condition of her own preferences. For example, Column might prefer icon the most but for Row a coordination on the icon might not be valuable. Row would select the reciprocated sign that satisfies her preferences over outcomes.

Chapter 8
Interpretations

There are two categories of interpretations. The first category consists of broad insights Nash equilibria reveal under alternative philosophies of mind. The second category explores and clarifies the type of explanations VSG equilibria generate.

The principal question is whether a place exists for mind in game theory. Rubinstein (1991, 923) affirmatively answers the question: "we are attracted to game theory because it deals with mind. Incorporating psychological elements which distinguish our minds from machines will make game theory even more exciting and certainly more meaningful." Rubinstein's claim relates to the philosophy of mind directly implying the following question: in what medium do Row's and Column's mental reasonings take place? Two answers to the question are available, one according to Cartesian dualism and the other according to monism that will be discussed shortly.

Common knowledge assumption of game theory implies that each player knows that the other can form signs, each knows that the other knows that each knows that the other can form signs and so on ad infinitum. Common knowledge of the ability to form signs constitutes players' consciousness of being involved in strategic interactions. It reflects a complex relationship between players' sensations, esthetic values, and ability to form signs that are pairs of "relata" (Barthes 1964, 655). Consciousness surfaces in players' esthetic abilities of evaluating how an abstract painting contains meanings for IR theory propositions and in players' interest in developing a common semiotic language in the Discipline. Players' mental states reflect their attitude and position toward theories and inexistence of a common language. Players are conscious of the value of coordination and the place of individual subjectivities in forming signs. Nash equilibrium proves to offer a rigorous language describing individual subjectivities and stability in interactive strategy choices.

Each equilibrium indicates proxemics, that is, social and personal distance between people and perceptions of this distance, implicitly and tacitly shapes human behavior and interaction (Hall 1966). IR scholars, whether they are structural realists or constructivists, positivists, and post positivists, have different ontological and

© Springer Nature Switzerland AG 2023
S. Ş. Güner, *Art and IR Theory*, Mathematics in Mind,
https://doi.org/10.1007/978-3-031-32342-3_8

epistemological attitudes. Proxemics yield alternative utility configurations. In a Saussurean game, for example, coordination can lead to different equilibrium payoffs as there exist commonality and differences in human understanding and experience (Schwartz-Shea 2006, 92). Differences in equilibrium payoffs ultimately relate to players' attitudes about what knowledge they can devise from constructed signs and whether this knowledge exists and contributes in the Discipline meaningfully.

VSGs incorporate two types of uncertainties on top of players' consciousness: like states which are constantly uncertain about reciprocal intentions according to Structural Realism, players interact under strategic uncertainty justifying the homomorphism assumption positing states as human beings. Constructivism does not claim a constant uncertainty of states, because states' intentions and perceptions of these intentions vary across anarchies. Players' uncertainty is constant in VSGs because no player is alone in making choices in Saussurean and Peircian games. Players must think about what choice the other will make under similar or different reasons of selecting or rejecting a sign under the constraint of strategic uncertainty.

Like states, players are helpless in detecting what states of mind others harbor. Row and Column cannot "have direct experiencing of mutual mental states" (Churchland 1984, 3). Thus, game theoretic assumption of common knowledge posits that players are informed of mutual preferences by sidestepping players' inability to experience what's going on inside their minds. Players assumed to be informed about the structure of a game, but they have to deal with strategic uncertainty to discover equilibrium strategies. An equilibrium occurring in strategy pair of (icon, icon) in a Peircian game means that players interact over utilities, not over mental origins of these utilities. Hence, players do their best to coordinate sign choices ignoring inner sensations of each other. Similarly, the multiple equilibria in a Saussurean game signal that Waltz's and Wendt's theoretical propositions are not identical across players' minds.

The central question of whether players' mental states and processes are to be found in their brains or they are completely independent of brain's physical domain demonstrates that the synthetic approach of VSGs emphasizing players' preferences over making meanings through strategic interactions exposes the connection between VSGs and the philosophy of mind. Thus, interpretations of VSG equilibria in both dualist and materialist philosophies of mind open paths to new discoveries. Evolutionary game theory constitutes an exception unless it is used to examine human interactions, since it is able to deal with, for example, interactions between flowers and bees and we ignore whether flowers or bees or both have conscious minds.

The philosophy of mind implies questions and poses problems to solve for artificial intelligence, psychology, communication theory, neuroscience, and cognitive sciences. Therefore, VSG analysis is also relevant for these latter fields. Two interacting computers mimicking players and searching for meanings of IR theory propositions on the basis of abstract paintings might not necessarily be a fantastic movie scenario or a dream. How do symbols sign equilibrium or equilibria point at affect and transform IR facts surrounding us? The question clarifies the problem of assessing a reality as if it is independent of players' thoughts, perceptions, and talk. The equilibria demonstrate structures that depend on scholars' minds and ideas. Coordination among scholars cannot be thought as based on correspondence between propositions about IR theories and their real-world referents as if the latter are independent from scholars' minds. This way of thinking makes us approach philosophical idealism according to which ideas are at the basis of human knowledge.

We are now ready to ask the following question: what is the nature of players' mental states in the condition of strategic interdependence? Game theoretic thinking answers this question by taking players as conscious of being part of a coordination process. To coordinate, Row and Column must ponder about mutual strategy selections and how to respond them. They interact according to their mental states covering their preferences over outcomes of their interaction and those over their actions. In a Saussurean game, if one thinks that the other would select R, then she cannot select A and if one thinks that the other would select A, then she cannot select R. Nash equilibrium concept rules these mental thought processes. In a Peircian game, players are conscious of the fact that they must coordinate over a sign and that a rejection of a sign is out of the question. Thus, Saussurean games allow more freedom to satisfy their subjectivities compared to Peircian games where players have to agree on a sign.

How do Row's and Column's mental evaluations get aligned (or not) on the meaning of a painting for a theoretical proposition? To answer this question, take, for example, Saussurean Stag Hunt game in strategic form at ordinal Level given in Fig. 4.18. Take the side of the Row and try to reconstruct how she reasons: "I really do prefer that I and Column we both agree that the Rothko painting signifies Waltz's proposition of structural constraints. There is no guarantee for mutual agreement to occur, however. Given the game matrix, it seems that Column could also select R. However, I cannot experience what mental state of Column is in. Column could satisfy her coordination need by selecting R to show her dissatisfaction with the painting by Rothko. Even if there exists no substitute for Rothko for the time being, her preference for coordination on an alternative painting might prompt her to choose R in the game. Against Column's choice of R, I have to select R too. We will miss the opportunity of obtaining the highest benefit of mutually consenting on the painting as a signifier. If I want to contribute to the common semiotic language, I have to align with Column's strategy because if (R, R) emerges as an equilibrium, we can perhaps at least agree on an alternative painting as the signifier. But for an agreement on an alternative painting to emerge as a signifier, Column has to align with my strategy of 'reject' for the same reason. In fact, I ignore why Column has such a preference ordering, why she values (R, R) more than (A, A). I cannot 'read' Column's mind about her preferences over the outcomes. I cannot be certain that while I work for the Pareto efficient outcome of (A, A), she can choose R. Column's choice of R against my choice of A means my worst dream. My best reply to Column's R choice is to select R as well." Column reasons in terms similar to those of Row. Thus, the philosophy of mind adds the condition of players knowing the meaning of a sign for themselves only to strategic uncertainty. Any sign coordination does not imply that players have identical mental states. It is impossible that a player can sense the subjective experiences of the other player.

Interpretations of Nash equilibria require the detection of elements intersecting game theory and doctrines of the philosophy of mind. They expose how game theory and philosophy of mind meet by divulging the intersection of mental phenomena and consciousness in strategic interactions. The intersections suggest alternative answers to the question of whether art can be vehicle for coordination among IR theorists on the principal propositions by Waltz and Wendt. Each doctrine implies alternative vocabulary, syntax, and semantics. Hence, each intersection demonstrates the existence of a multiplicity of truths encrusted in players' best replies across Saussurean and Peircian games and exposes a common language via art, game theory, and philosophy of mind.

8.1 The Place of Mind in IR

The problem of other minds is critical in analyses of international politics. Diplomatic history narratives ascribing beliefs, perception, and rational calculations to states demonstrate how mental factors are used by historians who reconstruct past events and interactions (Renouvin 1945; Gulick 1955). The following note by Gulick is instructive for an example among many others about states having human-like sensations: "While it was true that Britain had Hanover to worry about, England did not have to share her own territorial boundaries with another power in the continental sense; she could lose Hanover and remain untouched herself" Gulick (1955, 210).

Not only sources of diplomatic history but theoretical IR approaches as well consider mind as a central factor. Two main IR theories support this claim: Structural Realism and Constructivism. Both theories assume homomorphism according to which states are equivalent to humans. Humans have minds. Thus, the assumption implies that states have minds. The assumption reflects states' intentions, fears, security, and insecurity as nonphysical phenomena; they are not observed real objects. If mental activities are equivalent to consciousness, then states' views of each other and themselves are reflections of consciousness. Human mental processes and operations of intentions, fears, security, and insecurity become elements of states' consciousness in the context of interstate actions. For example, Waltz (1979, 105) remarks that:

> When faced with the possibility of cooperating for mutual gain, states that *feel* insecure must ask how the gain will be divided...Even the prospect of large absolute gains for both parties does not elicit their cooperation as long as each fears how the other will use its increased capabilities. Notice that the impediments to collaboration may not lie in the character and the immediate intention of either party. Instead, the condition of insecurity-at the least, the uncertainty of each about the other's future intentions and actions-works against their cooperation.[1]

---

[1] Italics mine.

Wendt (1992, 398) similarly remarks that:

> Each person has many identities linked to institutional roles, such as brother, son, teacher, and citizen. Similarly, a state may have multiple identities as "sovereign," "leader of the free world," "imperial power," and so on. The commitment to and the salience of particular identities vary, but each identity is inherently social definition of the actor grounded in the theories which actors collectively hold about themselves and one another and which constitute the structure of the social world.

Thus, both Waltz and Wendt support the assumption of homomorphism by ascribing human traits to states taken as individuals. The implication of the support affirms the place "mind" occupies in Structural Realism and Constructivism allowing, offering a space for the philosophy of mind in the Discipline. The end result is the relevance of the question of consciousness for IR theories.

In fact, Waltz is the IR theorist who explicitly recognizes the human mind as a central subject in the construction of theories. He defines theory as "a picture, mentally formed, of a bounded realm or domain of activity; a depiction of the organization of a domain and of the connections of its parts" (Waltz 1979, 8). The definition explicitly refers to the existence of mind-forming pictures as separate from the bounded realm of international politics. The dual aspect of the definition is due to a physicist. Waltz follows Austrian physicist Ludwig Boltzmann's neo-Kantian lead in his theory definition. Boltzmann (1974, 33) stated that the "task of theory consists in constructing a picture of the external world that exists purely internally and must be our guiding star in all thought and experiment." Waltz's theory definition follows Boltzmann's lead referring to a nonphysical mental picture referring to a relationship between human mind and the physical domain of states' interactions. Why does Waltz value Boltzmann's theory conceptualization? Because Waltz declares that "I consider myself a Kantian, not a positivist" (Halliday and Rosenberg 1998, 379). Waltz's preference to follow Boltzmann who relies on internal pictures of externality parallels Waltz's theory understanding of mental pictures of IR. Therefore, Waltz's theory definition implies the centrality of Neo-Kantian thinking generating the problem of how mind, a nonphysical substance, interacts with international politics corresponding to physical substance in space.

Structural realism and constructivism both rely on mental pictures. Wendt's proposition of dynamic nonunique anarchies constitutes a social picture of international politics incorporating social elements like states' intersubjectivity. States' understanding of each other based on their mutual perceptions and how each perceives mutual perceptions in turn creates a complex cognitive network seated upon the reality of international politics. The duality of abstractness of theory versus substantial international politics appears in a more complex fashion in Constructivism. Wendt proposes social elements as central forces in formations of alternative anarchies. Waltz's proposition implies a unique, unchanging principle of organization of international politics as being equivalent to the coexistence of sovereign units. It follows that the central difference between propositions by Waltz and Wendt is the relative abstractness of Waltz's theory compared to Wendt's. Waltz has a higher penchant for abstraction needing fewer abstract elements to describe the nonphysical substance of mental picture unlike Wendt.

To sum up, the philosophy of mind constitutes a powerful source of interpretation of diplomatic history and texts of IR theory. The philosophy of mind can enrich the Discipline through its multiplicity of branches studying mental processes. Dualist and materialist philosophies of mind constitute rich avenues to interpret IR theories in their dual and material forms. Each branch uncovers previously neglected features of both IR theories and historical narratives. The philosophy of mind is a valuable source of interpretations of equilibria computed in VSGs. Interpretations of VSG equilibria constitute a separate subject matter. But first we need to clarify the place of the philosophy of mind in abstract thoughts of Waltz and Wendt.

8.1.1 Dualism and Materialism

Blackburn (2016, 141) notes that "any view that postulates two kinds of thing in some domain is dualistic; contrasting views according to which there is only one kind of thing are 'monistic.'" The note offers a general definition of dualism. Fodor (1981, 114) offers a definition of dualism in the field of the philosophy of mind:

> Traditional philosophies of mind can be divided into two broad categories: dualist theories and materialist theories. In the dualist approach the mind is a nonphysical substance. In materialist theories the mental is not distinct from the physical; indeed, all mental states, properties, processes and operations are in principle identical with physical states, properties, processes and operations. Some materialists, known as behaviorists, maintain that all talk of mental causes can be eliminated from the language of psychology in favor of talk of environmental stimuli and behavioral responses. Other materialists, the identity theorists, contend that there are mental causes and that they are identical with neurophysiological events in the brain.

What type of duality does Waltz's definition imply? The following proposition by Descartes (2004, 41) echoes in Waltz's definition of theory as a mental picture of a bounded domain:

> Also, the fact that I find myself having mental images when I turn my attention to physical objects seems to imply that these objects really do exist. For, when I pay careful attention to what it is to have a mental image, it seems to me that it's just the application of my power of thought to a certain body which is immediately present to it and which must therefore exist. To clarify this, I'll examine the difference between having a mental image and having a pure understanding. When I have a mental image of a triangle, for example, I don't just understand that it is a figure bounded by three lines; I also "look at" the lines as though they were present to my mind's eye. And this is what I call having a mental image.

In Descartes' view, Waltz's mind is a nonphysical entity expanded in space. It creates a picture of a physical entity that is international politics. The result is the duality of nonphysical mental picture based on human consciousness and Waltz's power of thought in contrast with the physically observed realm of international politics. Hence, Waltz when he looks at international politics, he does not understand that international politics corresponds to relations among actors including states, international organizations, and so on. Rather, he "looks at" international politics as though they were present to the eye of Waltz's mind. Waltz's mind "sees" sovereign units that coexist and "counts" those states that are "superpowers" displaying a structure of the international system.

The duality Fodor refers to "monistic" view of materialism including identity theories that posit an equivalence of brain and mind. Structural realist theory of international politics is a mental image, that is, a mind product. Thus, in materialist philosophy of mind Structural Realism is nothing but a brain product. Structural Realism then becomes a research area of neural networks, brain research, and neurology. Whether IR theories are brain products constitutes a topic for materialist philosophies of mind that investigate whether consciousness emanates from brain. I do not explore consequences which follow the duality of brain versus mind but I explore consequences of the duality of mind versus international politics.

8.1.2 IR Theories as Mental Pictures

Cartesian duality constitutes a central topic in philosophy of science and philosophy of mind (Chalmers 1996; Churchland 1996, 1984; Dennett 1991). Waltz (1997, 913) indicates that as the molecular biologist Gunther Stent has put it: "Reality is constructed by the mind ... the recognition of structures is nothing else than the selective destruction of information (Stent 1973, E17)." If mind constructs reality, then Waltz's theory definition requires an answer to the question of how do mind and body interact.

The mind is the seat of consciousness generating pictures of the matter (Descartes 1989). According to Descartes' claim, the realm of senses constitutes a category distinct from the realm of reality; mind and matter are distinct substances which separately exist. For example, take the triangular relationship of China, Australia, and the United States. The relationship constitutes a physical realm. Following Waltz's theory definition, alternative theories including Constructivism can picture mental images of the relationship. One theory, for example, could be that of Georg Simmel who claims that all relations between three entities tend to evolve into an opposition between two against one (Simmel 1955). The theory would imply that any two states, say the United States and Australia form an alliance targeting China, China and Australia targeting the United States, and China and the United States targeting Australia. Any person without any IR theory knowledge would argue that the last two alliance configurations are not "meaningful"; the only meaningful configuration would be a United States-Australia alliance targeting China. The "meaningfulness" of the United States-Australia alliance stems from an unobservable individual reasoning. The question is then to detect reasons privileging the last configuration. Constructivist theory would answer the question as United States and Australia are "friends"; structural realist theory would answer the question as the need to balance Chinese power push the United States and Australia to align. Both answers are products of mind. Another theory could be the theory of structural balance proposed by Heider (1945) according to which relations are guided by the principles of "a friend of my friend, as well as an enemy of my enemy, is my friend; a friend of my enemy, as well as an enemy of my friend, is my enemy" (Antal et al. 2006; Auster 1980). Suppose that Structural Balance theory is used to assess the stability of the triad of Australia-China-United States. The theory implies that any alliance of two states against the third implies the stability of the triad. If a state, say Australia, has friendly relations with both the United States and China while the United States and China are enemies of each other, then no stability exists, because of Australia's inconsistent preferences. Australia's friend China is the enemy of the United States which is a friend of Australia. The configuration implies that my friend is the enemy of my other friend. The inconsistency is called "cognitive dissonance" (Festinger 1957; Goldgeier and Tetlock 2001; Ross and Ward 1995). The cognitive dissonance is supposed to generate problems and associated costs to deal with inconsistency and the violation of structural balance principles.

Waltz's approach to theory depends on Waltz's power of thought in words of René Descartes and the "plasticity" of his mind according to Churchland (1994). Mental pictures imply that states' intentions, fears, security, and insecurity require a theoretical approach; they cannot be understood by sheer observations. Hence, Waltz's mind functions as a mediatory device between abstractions and realities of international politics. Wendt (1995, 72) agrees with Waltz's remark: "Realists might point out that states can never be "100 percent certain" about each other's intentions because they cannot read each other's *minds* or be sure they will not change, and from this argue that since in an anarchy the costs of a mistake can be fatal states have no choice but to represent each other as enemies."[2] Wendt's nonunique dynamic anarchies and Waltz's constant unique anarchy reflect variable uncertainties among states but both propositions relying on states' uncertainty about each other imply that Waltz's and Wendt's mental pictures are not mutually disjoint.

Either Waltz's or Wendt's, human minds can conceive alternative constructs of international politics. The separation between Structural Realism and Constructivism emerges at the point of constructs such that in one type of construct a state's uncertainty is constantly high. In a realist anarchy states are uncertain about others' aggressive or nonaggressive intentions. Constructivist constructs permit states to be certain about mutual friendship in the international politics like in a Kantian anarchy or certain about mixtures of intentions of other states like in a Lockean anarchy. Realist anarchy as a principle of organization of international politics represents a fully uncertain environment as opposed to constructivist environments where such an uncertainty is reduced considerably.

Both types of constructs, be it realist or constructivist, indicate reasons Waltz and Wendt have for their claims of knowledge. Davidson (1963, 685) helps to formulate an answer to the question of how do mind and body interact in both theories:

> What is the relation between a reason and an action when the reason explains the action by giving the agent's reason for doing what he did? We may call such explanations *rationalizations*, and say that the reason "rationalizes the action…A reason rationalizes an action only

---

[2] Italics mine.

if it leads us to see something the agent saw, or thought he saw, in his action – some feature, consequence, or aspect of the action the agent wanted, desired, prized, held dear, thought dutiful, beneficial, obligatory, or agreeable. We cannot explain why someone did what he did simply by saying the particular action appealed to him; we must indicate what it was about the action that appealed."[3]

Davidson (1963, 686) further notes that:

> Giving the reason why an agent did something is often a matter of naming the pro-attitude (a) or the related belief (b) or both; let me call this pair the *primary reason* why the agent performed the action. Now it is possible to reformulate the claim that rationalizations are causal explanations, and give structure to the argument as well, by stating two theses about primary reasons:

1. For us to understand how a reason of any kind rationalizes an action it is necessary and sufficient that we see, at least, in essential outline, how to construct a primary reason.
2. The primary reason for an action is its cause.[4]

Davidson's arguments shed light on mental pictures of Waltz and Wendt when the arguments are transposed like in music theory from one key to another into mind-body interaction and imply that structural realist and constructivist constructs of international politics constitute rationalizations.

Both Waltz and Wendt have their own reasons to offer their theories. Waltz's primary reason lies in the structure concept as necessary to offer a sober version of Realism in his mind. Structural Realism cannot be formulated unless a mental picture is formed. Wendt's primary reason lies in the social structure approach as a necessary correction of Waltz's theory. Wendt believes that a relaxation of the abstraction level of Structural Realism is a necessity to understand the physical domain of international politics; not a unique and constant anarchy but nonunique dynamic anarchies describe physically observed realm of internationals politics. Constructivism cannot be formulated unless a more complex mental picture is formed. In both cases, both theorists' pro-attitudes and beliefs, that is, primary reasons cause their theoretical propositions. And what about the players interacting in VSGs? We now turn to find an answer to this question.

8.2 Dualism

A support to interpretations by an identification of commonality between game theory as a method and the philosophy of mind comes from Paul Feyerabend who indicates that "the interpretation of an observation language is determined by the theories which we use to explain what we observe, and it changes as soon as the theories change" (Feyerabend 2010, 219). Thus, the language of Nash equilibrium

---

[3] Italics in the original.

[4] Italics in the original.

changes as philosophy of mind theories used to interpret them change. To illustrate, dualist and materialist views allow reformulations and translations of Nash equilibrium language in their own language. Accordingly, Nash equilibria of Saussurean and Peircian games can take different meanings in terms of dualist and materialist views. To illustrate, the equilibrium of (A, A) connotes that the painting in question constitutes a signifier for a proposition that is the signified in different terms according to dualism and materialism, two principal schools of the philosophy of mind.

Cartesian dualism assumes two realms distinct from each other: an ordinary physical substance and a nonphysical substance that occupies no spatial position compared to the physical substance. Mental phenomena are nonphysical phenomena (Churchland 1984, 2). Mind, the nonphysical substance, thinks (Churchland 1984, 8). Players' mental states are related to their thoughts which are nonphysical events. Therefore, the nature of preferences over outcomes is nonphysical. Interpretations of Nash equilibria are to be conceived in nonphysical terms.

If one imagines how Row and Column think about how a painting constituting as a signifier for a signified proposition, then one has to admit that players' minds are not as real as observed paintings. Mental reasonings of Row and Column take place in the nonphysical domain for how paintings by Pollock and Rothko constitute real-world referents of Wendt's and Waltz's propositions. Consequently, Row's and Column's minds are strategically interdependent and interact. Sign making depends on players' minds. The question of: "where do our ordinary common-sense terms for the mental states get their meaning?" (Churchland 1985, 3) implies the question of "where do our ordinary common-sense term for the mental state of color interactions get their meaning?" in Rothko-Waltz semiotic connection. Similarly, the question transforms into "where do our ordinary common-sense term for the mental state of dynamic features of paintings get their meaning?" in Pollock-Wendt connection. Dualism answers both questions by pointing out players' minds as sources of making meanings. In short, perceptions of color interactions and dynamic features in paintings constitutive of meanings stem from inner mental states.

Dualism forces to ask and answer another question: "how can Row (Column) be sure that the inner sensation, to which Column (Row) has attached the term 'strong or weak color interaction in a Rothko painting' is qualitatively same as the inner sensation to which Row has attached that term of 'strong or weak color interaction'"? This is a semantical problem affirming that players have nonequivalent inner states. Hence, in a Saussurean game, Row and Column both choosing the strategy A or R, or Row and Column both choosing the same signs in a Peircian game emanate from different inner states of players. Coordination on the same sign does not imply that players have identical mental states. In fact, players cannot have direct experience of mutual mental states; Row cannot have direct experience of Column's mental state and vice versa in the process of making meanings. VSG players are both humans not of different species. It is difficult to imagine how a bat reasons or live through life (Nagel 1974). Yet it is still difficult for one player to understand how the other's mind works. If players' preferences over outcomes are explained and tabulated in a game matrix, then this does not imply that players experience mutual mental states like Row experiences Column's subjective views toward signs and vice versa. Thus, a coordination equilibrium does not have to be based upon equivalent mental states but differences in players' inner sensations do not hinder coordination necessarily.

Cartesian view of the existence of two forms of substance, namely, the substance that has no spatial position and the physical substance of the person, qualify the VSG equilibria as results of "nonspatial thinking substance" (Churchland 1985, 8). As the equilibria mean signs, dualism supports the view that players' physical strategy choices play no role in their formation. Physical selections of best replies follow preferences over outcomes in the nonphysical realm. Consequently, VSGs imply that signs are products of nonphysical thinking activities in situations of strategic interdependence where players are rational decision-makers. Strategic interdependence is equivalent to interdependent minds that are nonphysical, conscious, and rational intelligence.

Nash equilibria are the results of interactions between two substances having no position in the physical realm. No nonphysical substance can determine a Nash equilibrium alone; stability in interactions derives from interdependent substances having no position in space. A sole mind cannot settle the nature of the sign. Interdependent minds generate Nash equilibria.

The implication of dualism amounts to the claim of distinct conscious intelligences conducting similar calculations of best replies. The distinctness of Row and Column's conscious intelligences does not hinder or prevent similar or dissimilar reasonings in reaching Nash equilibria. Rubinstein's remark might be evaluated as supportive of dualism, but we will have a better opinion once we explore implications of materialist theories objecting against dualism. Another implication of dualism is that physical sciences cannot penetrate a substance occupying no physical position in space. Physical sciences cannot have an access to how Row and Column appreciate colors and interactions of hues in a Rothko painting or appreciating moves, places, and curls of color in a Pollock painting. These appreciations of players derive from players' introspections such as sensations, willingness to form signs, and fluctuations in thoughts about structural constraints and dynamic anarchies.

There are two doctrines that second dualist view of sensations as being out of reach by physical sciences. These doctrines are epiphenomenalism and interactionist property dualism. Both doctrines accept mental states in brain as a material domain and that physical sciences cannot explain sensations of conscious intelligences. Epiphenomenalism maintains that mental phenomena "do not have any causal effects on the physical world"; nonmaterial conscious intelligences cannot generate any causal effects (Churchland 1984, 11). It follows that players' sign choices cannot be interpreted as being physical choices while they are enacted by the brain. "Brain activities are causally impotent" (Churchland 1984, 12). If one maintains that sensations players experience out of paintings constitute mental phenomena termed as epiphenomena, then it is impossible to assert that these sensations have any effect on the equilibria and, as a consequence, on signs. Therefore, epiphenomenalism supports preferences over outcomes are alimented by sensations in brain. Yet it does not support the claim that preferences over outcomes shape behavior, that is, Nash equilibria constituted by best replies. Preferences over outcomes cannot lead to preferences over strategy choices and therefore Nash equilibria according to epiphenomenalism. Epiphenomenalism does not explain equilibria in short. In contrast, interactionist property dualism claims that sensations affect brain and behavior. It is not a materialist doctrine because it opposes the claim that sensations in brain constitute a domain which physical sciences can penetrate. Thus, interactionist view bends Cartesian dualism by its claim that color interactions in a Rothko painting or shifts and curls in Pollock painting affect brain. Brain constitutes the physical realm. Thus, first, players' introspections facing paintings prompt sensations generating preferences over the outcomes of interaction in brain. In the second step rational intelligent minds make choices in a situation of strategic interaction. Nash equilibria result from brain sensations, yet these sensations are not open to methods to study brain as a physical domain. Interactionist property dualism opposes epiphenomenalism by claiming that brain sensations can have a causal impact on players' behavior. Preferences are based on players' conscious intelligences and prompt preferences over actions on the basis of players' preferences over the game outcomes.

The materialist view of philosophy of mind opposes dualism in all its diversities. Instead of a dual nature of mind, materialism argues for a monist view claiming that brain is the seat of the mind. There is no dual nature of mind and body but a unitary view of physical substance of brain and body. The problem shifts to how to locate mind in the brain. The difficulty of searching for spiritual matter converts to searching for mind in brain, a physical matter. Yet neuroscientists' research on brain exposes the constitution of brain and how it physically works. Dualist response is that while brain is a special organ controlling behavior, it can only a placed between mind and body (Churchland 1985, 19). Brain mediates but does not generate sensations, qualia, or consciousness for dualists. To summarize, players' mental states in VSGs can be argued via dualism or materialism. Game theory taking preferences over outcomes as given but generating rigorous answers to preferences over actions is unaffected by dualism or materialism. It is possible to interpret game theory as a method to study interactions either between spiritual or physical matters in contexts of strategic interdependence and uncertainty. Players' mental alignments over signs are not to be found in game theory but in the philosophy of mind. And, finally, a coordination equilibrium can occur through players' mental evaluations identical or not. All players being distinct individuals it is not necessarily true that sign making is to be found in the same area their brain, and, similarly, It is not necessarily true that sign making is spiritually identical for all players.

8.3 Monism

Idealism and materialism constitute monist theories. Idealism posits that reality is completely mental. Nature does not play any role in players' preferences over outcomes and actions. No VSG occupies any physical position, all VSGs are purely mental. All VSGs happen mentally. Thus, idealism in this form cannot help to yield any interpretations of equilibria in Saussurean and Peircian games because Nash equilibria are observed choices taking place physically.

Materialist theories of mind are monist as well. They posit that the physical world is equivalent to the real world. They discard the concept of nonphysical mind concept disconnected from the real world. Mind cannot be an abstract object, a nonphysical entity. They oppose dualism and supports the view that dualism claiming a distinction of mind and body only complicates a simple issue. Mind is a matter, not a nonphysical substance. Hence, any Nash equilibrium of a VSG must be connected with the physical world, not with an abstract concept of mind.

Materialism does not exclude the possibility of computers instead of humans interacting in VSGs (Searle 1980). If this is the case, then VSGs become strategic interactions in a physical world with computer algorithms interact. Computers decide on what actions are best replies in VSGs. Do computers act through different or identical algorithms? If they run through identical or nonidentical algorithms, then would Nash equilibria eliminate one pair of best replies in favor of another one? On what bases would one decide that computers run the same or different programs? These become central questions of a materialist view of interacting computers replacing interacting human players. If there exists a possibility of running VSGs played by strategically interdependent computers, then there is the problem of whether programs or algorithms imply unique beliefs and desires forming physical mental states. Any choice of an algorithm over an alternative one presupposes that machines' "sensations" of relata are predetermined. How would a programmer decide that computer decisions over actions reflect identical or nonidentical "beliefs"? Players or computers might not hold qualitatively similar views about the nature of signs. Any preference for an algorithm would then constitute an arbitrary choice and would lead to automatically the same type of signs.

Here, materialism would argue that color interactions are felt in the brain, yet dualism would counter materialism as there is no undeniable proof of brain as the seat of color interactions qualia. Functionalism allows players' esthetic abilities and forming signs, depending on their individual tastes and subjectivities, and how formed signs at mental level shape preferences over outcomes of interactions. If one insists that one can search for the implications of the materialist view without proof, it is possible to enrich interpretations of Nash equilibria in both types of VSGs. Suppose that Row and Column communicate prior to their selections of strategies. They might inform each other how they feel about a drip-painting by Pollock. They can exchange views about swirls and turns all around the canvas and Row finds the painting as a symbol but Column as an icon. They have to face the reality of the game matrix, however. Either Row has to select icon or Column has to select symbol to reach a coordination equilibrium. The constraint of meeting minds means the constraint of selecting same strategies in the Peircian game. A similar thought experiment in a Saussurean game can be that while one prefers signifier nature of the Pollock painting for Wendt's proposition and the other not, either one has to change her strategy implying that one has to forgo her preferred sensations about the sign. Nash equilibrium functions as a constraint upon players' preferences over their actions and "betraying" their mental views including their qualia. The Nash equilibrium might not necessarily satisfy their sensations. Consequently, in functionalist terms, it is possible to claim that the subjectivity of the state of preferences over outcomes faces the objectivity of the state of Nash equilibrium conditions in VSGs.

Any similarity of the inner sensations of Row and Column is a chance event; not a certain one. How can Row (or Column) think that Column's (Row's) inner sensation of colors in a work by Rothko and swirls and curls in a Pollock drip painting generate iconicity (or indexity or symbolicity) of that painting for a theoretical proposition? Can Row or Column experience reciprocal inner sensations in sign making? The answer is simple: Row cannot experience Column's inner sensations and vice versa. Hence, players cannot penetrate reciprocal mental sensations prompting them to select a strategy in a VSG. Game theoretic assumption of complete information sidesteps the problem of players' mental sensing of similarities and differences in forming signs by simply supposing common knowledge of preferences and avoiding the problem. Common knowledge implies that each player knows, that is, is informed of preference orderings of others, yet common knowledge does not mean that each player knows how the other individually construct the same sign on the basis of identical mental states. Thus, the common knowledge assumption does not take note of players' impossibility of experiencing mutual sensations. The philosophy of mind takes note of that impossibility. If there is a coordination equilibrium, players' best replies become based on a supposition of complete information of preferences but not upon any information about players' mutual inner sensations either produced by players' brains or as nonphysical phenomena completely detached from material brain. As a result, if one conducts a statistical experiment asking, say, IR scholars about what they sense about a painting's iconness, indexity, or symbolicity for an IR proposition, it is false to argue that an experiment generating coordination emanates from scholars' identical inner sensations. Thus, a common semiotic language among IR scholars does not have its origin in scholars' identical inner sensations; equilibria of VSGs do not necessarily reflect identical mental processes. Materialist philosophies of mind oppose dualist doctrines by positing that inner sensations of mental states stem from the same brain region.

The last question is meaningful if two computers, instead of two human players were interacting in a VSG. The question then revolves around whether an interaction between two computers would create more insights compared to an interaction between two IR theorists. As Rubinstein (1991, 923) notes that the importance of game theory derives from its dealing with the mind. Hence, interaction between IR theorists remains more insightful unless machines have minds. If computers have genuine consciousness, then VSG equilibria might be equally insightful whether the interaction takes place between machines or humans.

Players' access to their own minds ultimately connects with interpersonal comparison of utilities because if a player has difficulties in "reading" her own mind, how could she read the other's mind and therefore the other's preferences over signs paintings generate for IR propositions of Waltz and Wendt? At this point game theory and the philosophy of mind concur opposing any idea of the possibility of interpersonal comparisons of players' preferences. From a game-theoretical stand-point the difficulty is avoided by preferences as given parameters of games, yet strategic uncertainty penetrating players' minds and strategic uncertainty given preferences of players constitute different subject matters.

Qualia corresponding to players' sensations, say when they gaze at a painting, help players form signs yet qualia remain beyond the reach of physical sciences. From this perspective of individuality of sensations, it is impossible to measure the intensity or what qualia correspond to. A Rothko or a Pollock painting generating sensations similar to having sensations of smelling a rose or having a pain are impossible to think and talk about in the absence of intelligence. Players' utilities are inexpressible outside players' minds, and therefore sources of their preferences over the outcomes in a VSG can be only assumed; they cannot be explained. Hence, the impossibility of justification or explanation of preferences is not surprising from a dualist perspective. The question of where preferences come from is "irreducible" (Churchland 1984, 12) as it is not answerable by physical sciences. Dualism supports game theoretical position of computing stability in interactions given preferences. Players, who are IR theorists, attach meanings to paintings and behave on sensations which cannot be explained in physical, electro-magnetic terms. Semiotics cannot derive any backing from physical sciences and remain anti-realist.

8.3.1 Behaviorism

Behaviorism rejects dualism. Gilbert Ryle (1949, 74) who started behaviorism claims that any difference between mental and brain states generates category mistakes that are presentation of things or facts of one kind as if they are elements of another one. The view of mind as separate from the physical realm constitutes the view of "ghosts in the machines." Ryle asks that "How can a mental process, such as willing, cause spatial movements like the movements of the tongue? How can a physical change in the optic nerve have among its effects a mind's perception of a flash of light?" (Ryle 1949, 19). Ryle's category mistake examples illustrate an objection against mental categories of willingness and perception as elements of physical categories of tongue movements and physical changes the optic nerve monitors. One category is presented as it belongs to the other in both examples. The problem remains to distinguish between physical and nonphysical mental states. Dualist interpretation of Nash equilibria transforms into best replies ghosts in the machines select in VSGs according to behaviorism.

Ryle has argued that one has to measure behavior and its causes. If one cannot observe nonphysical substance corresponding to mind and measure its effect upon behavior, then it is better to reject the dualist view. Therefore, players' intentions, stimuli, desires should be excluded in understanding VSGs and Nash equilibria. It is better to focus on Row's and Column's dispositions toward concocting signs for propositions by Waltz and Wendt through paintings. The central question becomes whether Row and Column are disposed to express their preferences by their strategy selections? Players should not be assumed as driven by nonphysical, unmeasurable forces but rather by their being disposed to select specific strategies leading to Nash equilibria.

Ryle's theory lays foundations for materialism. It draws its force from logical empiricism that proposed verification principle (Blackburn 2016, 281). The principle of verification implies that unobservable forces or events do not deem to be explored. Hence, mind as a nonspatial thinking substance does not make any sense unless it is observed and measured, that is, it is verified. In the process of verification, theoretical sentences are translated into measurable observed units through indicators and operationalized have an explanatory power; otherwise, they do explain or prove nothing. Therefore, an approach toward mind cannot rely on spiritual forces beyond the physical realm.

Behaviorism supports the view of behavior patterns getting rid of the spiritual component of dualism. Operationalization would imply the following: first, a player ascribing the meaning of index to a Pollock painting with respect to dynamic anarchies would insist that her view is correct; second, if she is given the freedom to select a painting to reflect what she feels about dynamic anarchies and selects a drip-painting canvas by Pollock, she would eliminate alternative paintings on the basis of her subjective feelings and qualia; and, third, she would insist on her view of indexity of the Pollock paintings, not on its iconicity or character of symbol in a Peircian game. Once all players' positions are observed, their behavior can be explained by equilibrium concepts depending on information and rationality conditions.

Behaviorism concentrating on observed interactions leaves out what subjective factors shape players' preferences. Thus, it reinforces game theoretical exclusion of subjective factors shaping preferences and supports the game theoretical position of explaining interactions given preferences. The exogenous nature of sources of preferences gets support.

8.3.2 Identity Theory

As a materialist doctrine, identity theory assumes that mind is brain. All mental states are equivalent to physical states of brain (Churchland 1984, 26).[5] Electrochemical mechanisms of brain trigger behavior. Thus, the human brain is the seat of all mental phenomena such as perception of color interactions in a Rothko colorpane painting and changing paths of alternative anarchies in a drip painting by Pollock. Brain's neurological networks are responsible for thoughts, pictures, and computations. Consequently, players' inner thoughts, views, and feelings of strategic interdependence are situated in brains of Row and Column. Nash equilibria, in

---

[5] Recent research proves how gut informs brain: How gut bacteria are controlling your brain – BBC Future

broad terms, become results of best replies computed by Row's and Column's brains. Preferences over outcomes and actions emanate from brain. In gist, minds interact according to dualism and brains interact according to identity theory. The problem becomes how do neural networks in brain produce preferences over outcomes and actions; are different signs emanate from different brain regions?

Identity theory posits that brain is the seat of the mind. It constitutes a materialist doctrine close to natural sciences. The implication of identity theory for VSGs is straightforward: Row and Column's mental states are physical brain states. Preferences of Row and Column originate from their brains. Hence, neuroscience constitutes a field to investigate the origins of Nash equilibria. For example, consistency of players' preferences over signs depends on neural networks brain contains. An immediate question is then whether different signs are connected with different brain regions and processes. How do neural networks of brain prompt players to prefer icon over index or symbol? An answer would be given by supposing that electrochemical neural networks of brain as the triggers of players' thoughts, preferences, and sensations.

Identity theory is an alternative materialist theory of mind. Its main claim is that all mental states are equivalent to physical brain states (Churchland 1984, 26). Players' brain states identify a painting as an icon, index, symbol, or a signifier for structural constraints and dynamic anarchies. The identification is a mental process involving brain leaves to room for a dual nature of the mind. To illustrate, subjective perceptions of color and color interactions a player senses facing a Rothko painting are nothing but some averages of "molecular kinetic energy" brain generates (Churchland 1984, 26). Hence, preferences are based on physical brain processes. Identity theory, unlike philosophical behaviorism, explains preferences instead of ruling out subjectivities as elements of a domain out of scientific reach. Causal mechanisms of VSGs leading to the icon-icon equilibrium are embedded in players' intersubjective beliefs and preferences over coordination successes and failures. Are these intersubjective elements observable? No, they are not. Harré (1988, 130) remarks that how brain's perception of pain messages when a dentist touches dental nerve of a patient must be described in detail as a causal phenomenon. He further remarks that "science deals with the observable. As nervous system's workings in the case of dental pulp cavity are not observable, one could interpret pain by not touching the dental nerve instead of injections of opium-based medicine found through scientific research over time" (Harré 1988, 131). Harré's remarks imply that electrical signals the mandibular nerve sends to brain must be exposed while they are not observable. This is a materialist view of philosophy of mind claiming that brain is the seat of pain evaluations through nervous system. Players in VSGs make mental calculations on the basis of their subjectivities on connections between art and IR sentences. The problem is whether these subjectivities come from the brain or not. The touch of the dentist on the dental pulp cavity corresponds to a flash of light in players' mind when they gaze at an abstract painting and connect these artworks as generating meanings for Waltz's and Wendt's propositions. Does sign making originate from brain or a nonphysical substance? Players' cultures can be argued to steer sign making but whether these sign-making processes stemming from cultures are material or not remains a central query. Grosso modo, materialist philosophy of mind supports the view that human brain is the seat of the mind (Churchland 1984, 2). Players' preferences are therefore to be found in players' brains.

The epistemological problem diverts attention to another aspect informing VSGs. Players who interact also assume that each possesses mental states and psychological moods. How do Row and Column know that reciprocal inner mental states exist? How can one demonstrate that Row and Column know that reciprocal preferences are connected to their mental states? Is it possible, for example for Row to know direct subjective experience of Column's inner state and vice versa? The assumption of common knowledge provides an answer: preferences are common knowledge; players do not need know how, which mental states are behind preferences.

Players' mutual ignorance about reciprocal mental experiences does not form a contradiction with the assumption of either complete or incomplete information. Game theory studies interactive decision making, not how minds work. VSGs assume that preferences are of common knowledge. Hence, even if a player or both players are uncertain about reciprocal preferences over outcomes of their interaction, they interact under the common knowledge assumption. Players' inner sensations are out of reach for them. Players know that coordination is better than no coordination. If the equilibrium occurs in the icon-icon, index-index, symbol, symbol pair, or the confirmation of signifier nature of a painting, then the equilibrium does not imply identical mental sensations. Players' inner sensations can be different radically, yet the difference does not preclude stability the equilibrium brings about. Therefore, the type of the sign the pair of the painting and theoretical proposition forming an equilibrium does not emanate from players' identical minds. The nonequivalence of players' minds does not preclude the occurrence of an equilibrium either. It is possible to imagine a case where players are exposed not only one but a multiple bi-color field Rothko painting containing different colors and investigate whether there is any coordination failure or not.

Players' mental processes and inner sensations take place in a medium that might be the brain or not. Materialist theories of mind defend the former view unlike dualist theories of mind that defend the latter one. A player's assessment of a painting as a signifier for or assign of structural constraints or dynamic anarchies reflects a brain process for materialism. Dualism instead argues that these assessments are nonphysical, that is, players' conscious intelligence is to be found in a nonphysical matter beyond the scope of physical sciences.

Row and Column's minds are distinct nonphysical elements according to dualism. The distinctness of minds entails that players evaluate semiotic connections between color-field paintings of Rothko and structural constraints and those connections between drip-paint canvases of Pollock and dynamic anarchies differently from each other. Any equilibrium reflects players' best replies yet based on players' distinct appreciations of signs. Hence, Row and Column are not identical in mental terms yet able to coordinate actions on different grounds. With brains only possessing special properties that are fundamentally distinct compared to other human organs, Row and Column can feel different sensations of colors, colors' placements and moves gazing paintings by Pollock and Rothko. They create meanings in distinct mental terms. Therefore, the equilibria represent coordination as stable outcome conscious intelligences entail. Is there any way to combine these arguments in a single philosophy of mind? My answer is yes, such a combination can be realized under functionalism.

8.3.3 Functionalism

I claim that functionalism permits a philosophy of mind compatible with the hybrid method of VSGs. First of all, functionalism is agnostic about mind versus matter. It does not accept the duality and materialist, physicalist approaches. It proposes that mental states are "triplets of relations: what typically causes them, what effects they have on other mental states, and what effects they have on the behaviour" (Blackburn 2016, 192). Thus, functionalism permits mental events being nonphysical matters and also allows for Nash equilibria as the behavioral component of players' mental states. Functionalism expands behaviorism's claim of environmental inputs yielding behavioral outputs by proposing networks of mental states affecting each other and behavior. It also rejects identity theory view of a single type of a mental state due to a single brain state. Functionalism supports the view of Row and Column not possessing identical mental states in proposing signs unlike identity theory positing equivalent mental states due to identical physical brain states of Row and Column. For example, Row seeing red and green Column might see blue and orange so that they see complementarity of colors, but colors seen can well be different.

Functionalist philosophy of mind implies that VSGs stem from four classes of mental states:

1. Players' interest in developing a semiotic language in the Discipline given the lack of common language.
2. Players' esthetic abilities and forming signs, depending on their individual tastes and subjectivities.
3. Players' preferences over outcomes of interactions.
4. Players' preferences over actions. Each class of mental state has causal roles. The first two classes shape the class that in turn second one. Preferences over outcomes determine preferences over actions and therefore Nash equilibria. The assumption of several mental states in the form of consciousness describes a space of reasons and corresponds to a functionalist view of mind (Sellars, 1956). It rejects materialist and physicalist views of mind according to which physical reality does not allow any place for nonphysical matters.

These four mental states drive VSGs in the context of a functionalist philosophy of mind: first, players' interest in developing a common semiotic language; second, players' esthetic abilities and forming signs depending on their individual tastes and subjectivities; third, players' preferences over outcomes of interactions; and fourth, players' preferences over their actions. Players' mental states derive from various factors like the need of developing a common language, interactions among IR scholars through various channels, the Discipline's neglect of semiotics, the lack of explorations of structural constraints and dynamic anarchies, and how each of these factors shapes players' preferences and therefore their strategic behavior. The mental state 1 shapes the mental state 2 that in turn shapes the mental state 3 that shapes players' preferences over their selections of strategies and action and therefore Nash equilibria.

The four mental states summarize strategic interactions over mentally offered propositions through their players' subjective interpretations. For example, players can select an alternative painting to form a sign by selecting the strategy R, or A in a Saussurean game as players' inner thoughts permit mental formation of signs through pairs of abstract art and theoretical propositions. According to the functionalist view, VSGs do not imply interactions as flowing out of material sources like players' brains as, for example, identity theory would assume. Both players' different mental states can align about the existence of the color interaction. Hence, if Row senses an intense color interaction in the "Green and Tangerine on Red," but colors are inverted for Row and Column, it is still possible that the painting constitutes an icon for Waltz's proposition in Peircian terms and a signifier in Saussurean terms. Color inversion not affecting Row and Column's mental states leads to a coordination equilibrium as a result. Thus, it is not important what type color perceptions players develop in their mental states for reaching a coordination equilibrium in a VSG. Interpersonal comparisons of color sensations do not matter. If, no color interaction, no sign can form. To sum up, players recognize patterns in paintings and connect them to structural constraints and dynamic anarchies. Players can argue that either the painting in question does not mean anything and an alternative painting is better for the same purpose, or they can arrive at a mutual understanding.

The mind-body problem implies that players' minds correspond to players' non-physical conscious intelligences which are separate from their brains. The answers will help to assess what the sort of evidence is needed to argue for the usefulness of a theory of mind to interpret game rules and equilibria. Take, for example, a drip-paint canvas of Jackson Pollock. An interpretation of colored curls and swirls by Row and Column refers to how their mind associate the meaning of changing ideas of states and therefore intersubjective cultures among them alter over time. The interpretation simply overrules Waltz's idea of the unicity of anarchy and the non-unicity of anarchies due to changing mental evaluations. These evaluations can be assessed by Row and Column not by going to an art museum and glancing a drip painting by Pollock but simply doing an internet search and finding the painting in virtual reality. Both types of evaluations transforming into interpretations through minds of Row and Column can be qualified as real (Chalmers 1996).

VSGs affirm that each individual is unique, complex, has her own feelings and taste. Players acknowledge that each has intuitions, feelings, and instincts. What one player believes about the other player's belief and conscious experience or what does one player know about what the other player knows on reciprocal sensations becomes embedded in the assumed condition of common knowledge. Players' attribution of meanings to paintings, say, like a Rothko painting constituting an icon for Waltz's proposition of structural constraints, deserves to be interpreted in both materialist and dualist philosophy of mind terms.

In functionalist terms, if players perceive color interactions in a Rothko painting and identify them as making the painting as an icon of structural constraints and they agree on that sign in a game, then this means that the icon-icon equilibrium is reached with players having no clue on how each other forms mental pictures individually. Row and Column, being able to dwell into own consciousnesses and being informed of mutual preferences, cannot prove exact sources of the other player's mental work. Row and Column cannot "have direct experience of mutual mental states" (Churchland 1984, 3). Consequently, a player being informed of game rules including the other player's preferences is not aware what happens in the other player's mind and how she that preference ordering. No player needs to have that information to assess what her best reply should be, because they act on the basis of given preferences. No player needs be informed of how the other player's mind functions. Yet, a closer look about these mental processes is helpful as game theory is a method of how to assess preferences over actions, not a method how preferences form; it needs no information whatsoever about players' preferences over outcomes. Game theory leaves out the question of how players' preferences and utilities as quantification of preferences form. Luce and Raiffa (1957, 12) promptly informs that "utility theory is not a part of game theory." Game theory simply serves to compute best replies given preferences. Therefore, different schools of the philosophy of mind shed light in alternative ways on the problem of the formation and the meaning of players' preferences game theory does not dwell into.

Qualia, for example, become meaningless according to behaviorism but not so according to functionalism. A closer look at color interactions is instructive. Albers (2013, 1) maintains that "a color evokes innumerable readings." Row and Column do not have to be experts in color theory, they do not have to be scientists having the knowledge of wavelength of light and color. One cannot either expect that all IR theorists are able to distinguish between different color intensities and contrasts in hues of color.

Color sensations are qualified as "qualia" (Churchland 1984, 13). In epiphenomenal terms, qualia have no effects upon behavior. In functionalist terms, qualia as mental phenomena can shape players' preferences over the outcomes which is a mental state as well. There is a connection between one mental state and another mental state. For example, Row and Column can experience qualia standing in front of a Pollock or a Rothko painting. They can experience qualia contemplating these artworks and they can experience them differently. It is impossible to measure the intensity of qualia in the form of feelings and detect their origins. Differences in mental processes yield differences of sensations color interactions. Yet players can experience different color interactions prompting to assert that a Rothko painting signifies Waltz's structural constraints, the signified, or that the painting is an icon of Waltz's structural constraints. Similarly, players facing a drip painting by Pollock can sense that Wendt's dynamic nonunique anarchies or Wendt's anarchy classification are embedded in the canvas. Do brains of Row and Column engender these

qualia? Are they neural activations? Perhaps. There is no proof of neurological activations in brain engendering sensations, however. Consequently, sign-making processes emanate from these inexpressible (and inexistent experiences according to
materialism) of Row and Column.

Row and Column can feel "color relatedness" in a color-pane canvas of Rothko
under their nonphysical mental states. Players, "even if they have untrained eyes
will recognize mixtures of colors" (Albers 2010, 37). Hence, Row and Column are
to harbor mental processes as being sensitive to color. If one asks on what grounds
can players assume that each enjoy any mental states, then this is the answer. Simply,
they are human beings. The quantity, the placement, and the intensity of the color
space generate innumerable color interactions which Row and Column can feel.
Row and Column can connect color interactions with structural constraints arriving
at relata of interacting units and structure of international systems in terms of arrows
in Waltz's figure. Waltz does not specify the intensity of the interaction between
structure and interacting units and therefore the intensity of structural constraints
unlike a Rothko canvas does. Row and Column, however, can sense the intensity of
structural constraints through color interactions. Hence, sensations enrich structural
constraints ascribing them intensity.

Naturally, players' sensing of color interactions does not have to be of the same
intensity; Albers (2010, 3) notes: "If one says "Red" (the name of a color) and there
are 50 people listening, it can be expected that there will be 50 reds in their *minds*.
And one can be sure that all these reds will be very different." Row and Column
looking at the same canvas by Rothko can sense color interactions, but they do not
have to be equally intensive. Therefore, looking at the same painting of Rothko both
Row and Column can perceive an intensity of color interaction, but their qualia do
not have to be of an equivalent magnitude. Still, they will be able to propose a sign.
In the Peircian game, for example, Row's and Column's evaluations of a Rothko
painting in question can lead to the equilibrium of icon based on different or similar
sensations of interaction. Players' choice of index and symbol necessitates more
sensual fine tuning compared to icon as similarity can be constructed more easily
than index and symbol. An icon is more visible than an index and symbol in short.
To illustrate, the canvas "Green and Tangerine on Red" by Rothko cites three colors
yet each color connotes innumerable greens, tangerines, and green-tangerine interaction on innumerable reds for the players. Albers (2010, 33) defines "optical mixture" as two colors engendering a third color by being simultaneously perceived.
Two colors on the painting combine and a new color emerges. The new color then
indicates structural constraints Waltz refers to through a Rothko painting. The new
color can vary in terms of its brightness and lightness, and, consequently, "seeing
what happens between colors" defined as interaction color by Albers (2010, 5) constitutes a subjectivity connoting innumerable qualia players' mental processes can
connect with structural constraints concept of Waltz.

Goodman (1968, 46) posits that the word "expression" refers to feelings, but the
word "representation" refers to objects or events. Goodman's view accentuates
players' evaluations of Pollock and Rothko paintings as representing Waltz's and
Wendt's propositions but players can feel that while Pollock expresses dynamism,

Rothko expresses inertia, respectively. Two IR theorists looking at paintings can base their sign propositions on things the painting represents, properties the painting possesses, and feelings the painting expresses (Goodman 1968, 50). Thus, expressions of feelings as qualia take infinitely many values. The endless range of intersubjectivity in VSGs in the sense of qualia means that, concerning a drip-painting by Pollock, players sense curls and color interactions possess an endless nature of change. Consequently, Pollock's drip paintings possess more direct figurative expression than a literal conceptual representation of dynamic anarchy. Words become weaker compared to the intensity and the richness of qualia players experience.

The complexity of players' evaluations of the paintings and associating them with the propositions by Waltz and Wendt contain further complexities. For example, within the functionalist approach to mind, players' esthetic and subjective mental states can have four variants in art (Audi 2015, 348):

1. The first variant is the direct transmission of a painter's feelings to the perceiver. Therefore, IR theorists receive Rothko and Pollock's feelings gazing at Rothko and Pollock's paintings. Players use Rothko's and Pollock's feelings they perceive in the preplay encounters to claim that the artwork in question corresponds to either icon, index, or symbol of anarchy and structural constraints. The variant is utterly subjective. There are feelings of the artists, these are directly transmitted to the players, and the players use the transferred feelings in the preplay discussions.

2. The second variant corresponds to the expression being equivalent to intuition that is both in the mind and in the artwork (Croce). The game displays interactions between players' intuitions alimented by artists' intuitions. A player can propose, in mental and intuitive bases, that a Pollock painting constitutes an icon, index, or symbol with respect to Wendt's anarchy concept. If both players agree on one of these signs, they reach a point of coordination rewarding them mentally and intuitively.

3. The third variant focuses on the artist's mind only excluding the viewer's feelings. The painting is an endpoint of clarification, articulation of undefined feelings at first. Croce extends this notion to everyone asserting that we are all artists. Croce (1965, 8) supports the view that there exist nonverbal expressions. Pollock and Rothko express their feelings and emotions through forms they concoct proving that they see things nonartists do not. Conversely, players being no artists see what the painters do not see in their paintings because they use them to make signs for IR theoretical statements the painters are unaware of. An immediate implication of this remark is that sign making in this context on the basis of paintings is "a product of brain, not a product of hands" (Croce 1965, 10). Sign making is therefore a mental process. Collingwood shares the same view. If players discuss before they make their strategy choices and exchange their feelings about the painting and its connection with an IR theory statement, can change, revise their positions during the process transforming the game structure and therefore the equilibria.

4. The fourth variant is the view of Susanne Langer. It presumes that feelings and emotions are expressed through structures and forms. Rothko's paintings contain parallel arrangements of two or more color-fields attracting the gaze of viewers in a hypnotic fashion, Pollock's paintings have forms revealed by drip painting with spontaneity and no particular directions. Players discuss what signs they make of these forms and try to coordinate their choices.

In summary, functionalism, as an alternative materialist view, affirms the existence of relations among nonunique mental states. Thus, not players' brains but different mental states of players generate preferences over outcomes of the interaction and over actions. Any equilibrium is nothing but a result of alternative mental states with respect to the coordination. Functionalism therefore offers a complex view of the mind of players by indicating not a single but a multiplicity of sources affecting players' interactions. Players' mental states can be traced back to interactions with different mental states again located in brain. Qualia are out of the picture as, for example, color interactions even if colors are inverted are matters of how different mental states interact. Players' preferences, for example, cannot be argued to have an identical origin across Row and Column and that identical origin being located in the same region of Row and Column's brains. Therefore, functionalism supports the view that players do not possess the same type of mental state gazing art. Row and Column cannot experience necessarily equivalent mental structures and activities. Row and Column cannot experience the same feelings and perceptions due to differences of interactions among different mental states they harbor. Identity theory would imply, contra functionalism, an easier justification for players' behavior pointing out an identical source meaning an equivalence of unique mental states while proposing signs. Now we turn to the second category of interpretations.

8.4   Explanations Versus Nash Equilibria

VSGs are not physical science experiments. They are hypothetical strategic interactions between players who are conscious beings. Players have such subjective traits as intentions, purposes, and interests which are not easily observable or measurable. Collingwood (1994, 213) remarks that human subjectivities constitute matters to which experiments and observations cannot have an access. What type of explanations VSGs offer under these subjectivities?

8.4.1   Explanations Why

Do VSGs explain coordination in sign making given players' subjectivities? Empiricism rejects unobservable knowledge altogether; according to empiricism, knowledge and explanation are equivalent to observations. If knowledge does not derive from observed experiences, say, laboratory experiments of VSGs, then Nash equilibria explain nothing. VSGs are useful tools to generate knowledge under the condition of the possibility of VSG experiments and observations. No observation, no knowledge in short.

It is impossible to deny the importance of observed communications between Row and Column in interaction. Empiricism would value such observations of communication as constituting a precious source of information and insights for VSGs. Yet observations of players' communication present formidable complexities: a closer look at players' communication necessitates an exposition, a categorization of messages, perhaps an infinite, unknown number of speech-act messages going in between an utterer and a listener (Sally, 2002). How to sift through complex communication data to arrive at messages constituting origins of Nash equilibria? What sort of data would reveal subjective origins of best replies? There exists a multitude of conceptualization and operationalization of observations as counting for and connected with Nash equilibria or no equilibria. Therefore, it is difficult to answer these questions.

A second empiricist problem is the theory-ladenness of data, that is, data preventing ultimate and unchanging truth as it is selected and used across experiments differently. To illustrate the problem, suppose that a structural realist and a constructivist are invited to interact in a VSG experiment. Each player would perceive games differently depending on abstract thinking covering both art and IR theories. Each is able to evaluate signs differently as theoretical lenses they count on shape their esthetic and abstract views.[6] Observations of Nash equilibria indicating best replies derive ultimately from players' preferences over outcomes shaped by players' subjectivities which are not observed. Empiricist view of the equivalence of knowledge and observations implies that ultimate and unobserved subjective sources of Nash equilibria are not considered. Bayesian VSGs count for such intricacies and resulting equilibria, yet the problem of theory ladenness remains. Einstein's dictum quoted in French (2016, 3) as "If the facts do not fit the theory, change the facts" constitutes an exit from the problem as long as facts experiments dwell into can change. Nevertheless, the dictum does not serve empiricism. The terms of theory and explanation irritate empiricists (Godfrey-Smith 2003).

Scientific realism versus constructive empiricism (Musgrave 1985) is instructive about what VSGs can explain. Constructive empiricism proposed by Van Fraassen (1980) implies that there is no guarantee of Nash equilibria observed in experiments constitute explanations. Constructive empiricism would take a successful explanatory theory as "empirically adequate." Van Fraassen (1980, 12) describes scientific realism as "science aims to give us, in its theories, a literally true story of what the world is like; and acceptance of a scientific theory involves the belief that it is true." He then claims that explanations of a successful theory are empirically adequate only; they should be taken with a grain of salt. Hence, constructive empiricism

---

[6]An evolutionary VSG positing players who espouse structural realist and constructivist views of international politics selecting A and R strategies in conformity with their theoretical lenses in a Saussurean game would permit to study sign making leading to polymorphic ESSs.

imposes the condition that Nash equilibrium explanations of coordination must expose unobserved features such as those subjective factors lying beneath players' preferences over outcomes and over actions. These unobserved features do not exist independently of human thoughts and perceptions. They cannot be directly described as underlying causes of preferences over outcomes or actions. VSGs are not empirical or theoretical devices; they do not imply successful explanations as implied by scientific theories.

Scientific realism has an opposing viewpoint compared with constructive empiricism. It can deal with unobservable mental entities. It accepts that mental explanations, if successful, are true. Thus, scientific realism would not discard Nash equilibria as empirically adequate explanations only. Consciousness may not be observed but this not a problem for scientific realism unlike constructive empiricism. Does consciousness exist? Scientific realism answers the question in an affirmative mode under one condition: one must explain coordination in VSGs made possible by consciousness. For example, the question of the possibility of an "icon, icon" equilibrium in a Peircian game can be taken as a scientific realist query. Game rules exist including players, players' preferences over outcomes and strategic interdependence. If one explains Nash equilibria on the basis of unobserved mental reasoning, then consciousness exists in players' mental calculations in strategic interactions. A parallel example can be constructed in structural realist terms such that Waltz's concept of structural constraints explaining states' alignment choices. Even if structural constraints are inexistent but enable how or why a state align to balance or to bandwagon, then explanation is successful and infers that structural constraints exist in reality. One has to describe the role of structural constraints through explanations, simply. Yet, there is no successful explanation of how structural constraints shape and shove states' balancing and bandwagoning alignment choices. VSGs generate a hint about structural constraints on the basis of semiotics, players' making meanings through art. In specific terms, players "observe" or not those constraints in a Rothko painting. The painting then becomes a medium, an intermediary communication channel between players. Players' senses of affirming structural constraints cannot be observed but a Nash equilibrium of (A, A) in a Saussurean game supports these subjectivities. Game theory then serves as a tool of disciplining players' interactions through unobserved mental features and help to generate explanations through equilibria. Scientific realism and constructive empiricism require theories that explain. The hybrid method is not a theory. Therefore, both philosophy of science doctrines are not useful to interpret Nash equilibria as successful explanations.

What does an explanation mean? Wendt (1998, 104) remarks that "merely descriptions of mechanisms generating facts are associated with "how" questions are scientific realist; deductions from laws of nature associated with "why" questions are logical empiricist." The remark succinctly summarizes Van Fraassen's "empirical adequacy" argument and scientific realist preference of avoiding the fulfillment of the observability condition underlying successful explanations. What does "explanation" mean regardless scientific realism and constructive empiricism? One answer comes from the Covering Law Theory of Explanation proposed by

Hempel and Oppenheim (1948) known as theory of nomic explanation. The theory claims that an explanation is a demonstration of how explanandum derives from explanans. The explanandum contains a law and some premises, that is, an observed regularity. In the Discipline, there exist no general law obtained under history-long observations such as weak states aligning and forming alliances to balance strong states. Hence, it is impossible to explain the formation of an alliance of weak states balancing the strong state in an international system using Hempel and Oppenheim explanation model that requires a general law. There exist alternative explanation formats criticizing, supplanting, or revising deductive-nomological approach in physical sciences (Pitt 1988). None of these is compatible with the hybrid method of VSGs.

Nevertheless, it is worth to ask the following question: what do nomic explanations teach in VSGs? First, VSGs are not fields of inquiry in the realm of physical sciences. Second, there exist no general regularities established as laws in VSGs. There exist no observed regularities of Nash equilibria in games either. Therefore, one cannot explain and interpret an equilibrium of a VSG as if game rules constituting "explanans" and equilibrium strategies as "explananda." Nash equilibria do not result from an established regularity; thus, they cannot constitute answers to "why" questions Hempel and Oppenheim pose because they do not follow any natural law to use as an explanans. Nash equilibria deriving from players' preferences over their actions constituting mental interactions are not "nomic necessities" (Cross 1991, 237). They do not constitute a deduction from an observed regularity like "all metals expand when heated, iron is a metal, therefore the iron strip under observation expands when heated." There exists no similar lawlike regularity in strategic interactions. There exist no correspondence rules connecting theoretical concepts of VSGs with empirical phenomena logical empiricism necessitates either (Godfrey-Smith 2003, 35).

In summary, Nash equilibria of VSGs do not explain sign making through coordination in the real world. VSGs do not aim at an overt research for regularities in equilibria, testing hypotheses based on empirical data of strategic interactions VSGs assume. VSG equilibria are not causal explanations in the sense of game rules and players' consistent preference orderings come first (explanans) and they determine equilibrium strategies in observed real strategic interactions (explanandum). They classify some strategy profiles as stable as no player has any incentive to deviate from her equilibrium strategy given that the other sticks to her equilibrium strategy. The classification indicates that there exists no "temporal asymmetry" in Nash equilibria (Wendt 1998, 105). There is no counterfactual condition of explanations in VSGs either because there is no temporal asymmetry (Wendt 1998, 105).

While scientific realism positively approaches the place of players' minds in VSGs under the condition of generation of successful explanations, scientific idealism posits that reality depends on mind (Audi 1999, 412). A painting by Rothko or by Pollock is a real object, an element of the external world. The meaning the painting generates for Row and Column reflects how Row and Column's minds operate. Nash equilibria then become a product of interdependent minds through strategic uncertainty. Consequently, under idealism it is possible to interpret signs players agree upon like icon, index, symbol, or their existence as reflections of individual but interdependent minds.

Returning to the possibility of VSGs run in experiments, we can note that "an adequate *explanation* of the real always requires some recourse to the operations of mind" (Audi 2015, 492).[7] There are no signs without minds according to idealism. Were players interacting for monetary profit, the reality in material terms can be argued to exist. Players interact over nonmaterial meanings not over material gains in VSGs. VSGs constitute a mediation between players' minds and constructed realities of signs only. They open a door to cognitive access in interdependent sign-making processes. If someone "smells" the scent of apple flowers by looking at the Mondrian painting "Apple Tree, Blossoming," it is impossible to exclude how players' minds shape their confirmation of the sign the painting forms for a theoretical proposition. Thus, meanings of artworks change and depend upon minds; they are not sitting there in reality to be discovered independent of players' minds.

8.4.2 Explanations How Possible

There exists an alternative interpretation of Nash equilibria of VSGs: they can be interpreted as answers to the question of "how possible?" (Dray 1964, 19). For example, how possible is that strategy profiles *(R, R)* or *(A, A)* constitute equilibria in a Saussurean game? The answer must expose players' mental calculations and the game rules leading to players' best replies. The equilibria are not used to understand some particular real-world interactions between IR scholars who interpret theoretical propositions via abstract paintings of Pollock and Rothko. The Nash equilibria do not explain why players chose equilibrium strategies in reality but answer the following question "how could it be that a certain thing happened"? (Dray 1968, 390), that is, "how could it be that rejection against rejection and accept against accept constitute best replies against each other?" The question in the context of a Peircian game could be posed as how could it be that symbol-symbol equilibrium be reached in the light of players' not concurring fully on symbol-symbol equilibrium? Players' objective of coordination under the constraint of strategic interdependence and uncertainty answers the question.

Rappaport (1995, 430) affirms that "If their explanatory force is not in general derived from the concepts incorporating general laws, what does make explanations by concept explanatory? My suggestion is simply that such explanations *unify* a range of events or phenomena under a *single concept*. It is by providing a unification of otherwise disparate phenomena that an explanation-what achieves explanatory power."[8] Accordingly, the Nash equilibria of VSGs unify a range of events or phenomena under the concept of best replies. The single concept is Nash equilibrium

---

[7] Italics in the original.
[8] Italics in the original.

that indicate players' best replies across a variety of VSGs. Thus, explanations by the "concept" of Nash equilibrium amount to interpretations of stability players reach in VSGs (Rappaport 1995, 424). The social structure and game rules summarize properties of VSGs; it is not used to detect lawlike generalizations. Social structure of VSGs summarized in players' preferences together with game rules such as players' information conditions and sets of strategies correspond to players' interdependent reasonings that follow players' mental calculations. The dependence of players' interactions on their minds make VSGs "theoretical models" (Rappaport 1995, 423).

Utility theory is independent of game theory. It must reveal sources of players' preferences. The sources of preferences are open to insights to gain from mental reasons. Davidson (1963, 685) argues that an agent's reason causes the agent's action. The reason constitutes the agent's justification of what she did. The justification derives from subjective elements such as the agent's desires, needs, wishes, cravings, longing, and fancy. One has to clarify why those elements explain an agent's action based on the agent's reason. Therefore, how possible explanations correspond to explanations in VSGs because they clarify causal connections between players' reasons to select, accept, or reject signs.

In more specific terms, Davidson's claim is based upon the concept of "primary reason" that is a pair composed by agents' pro attitude and beliefs and that is the primary reason for choices of actions (Davidson 1963, 686). Agents' pro attitudes and beliefs serve to unify different stability configurations under the concept of Nash equilibrium. Players have an intention to coordinate because they have the pro attitude to reach a common semiotic language due to the conflictive nature of the Discipline. Yet, the pro attitude alone is not a sufficient condition for an equilibrium. They must reason, say in a Saussurean game, like "if I choose signifier it is possible that the other chooses not signifier...if I choose non-signifier, then it is possible that the other chooses signifier." Thus, a Nash equilibrium has two sources: its quality of being a concept unifying stability conditions across VSGs and its origins in players' reasons and beliefs. Players' reasoning generates an answer to the question of how possible. The unified concept of Nash equilibrium in terms of Dray becomes directly related to Davidson's claim of players' reasons and beliefs generating Nash equilibrium.

Davidson's claim implies that it is necessary to describe primary reasons of Row and Column corresponding to their beliefs and pro attitudes of Row and Column beliefs regarding strategy choices in Saussurean and Peircian games. In a VSG, an agent selects a specific strategy because of her need for coordination, common language, and her aesthetic values in forming signs of IR propositions through paintings. The need for coordination, common language and individual aesthetic values form appealing elements of strategy choices under strategic interdependence and uncertainty. Hence, only one player's primary reasoning is not decisive in coordination: both players' primary reasonings incorporating their individual pro attitudes and beliefs and also their strategic interdependence shape equilibria of VSGs. If both players desire coordination and believe that a Nash equilibrium action leads to coordination, then Row and Column will select best replies in a Saussurean or a Peircian game.

Reasons at mental level cause actions. Davidson's thesis implies that the operation of a reason constitutes a mental event (Blackburn 2016, 405). Hence, preferences over outcomes reflect players' states of mind. Players confront the problem of assessing how their co-players' minds work in comparing outcomes' subjective values relative to each other. In a Saussurean game, for example, players can be both attracted to the *(R, R)* outcome sensing that either the painting in question is meaningless to form a sign or simply that no art is useful to discuss propositions of Waltz and Wendt. Preferences explain rationalized, not rational strategy choices of players when the outcome of interactions depend on both players' actions. If one explains why Row or Column or both choose a strategy in a Saussurean or a Peircian VSG by giving players' reasons for choosing, say, the strategy of signifier, then the explanation is called "rationalization"; "reason *rationalizes* the action" (Davidson 1963, 685).[9] It is impossible to explain an action "why someone did what he did simply by saying the particular action appealed to him; we must indicate what is it about the action that appealed" (Davidson 1963, 685). Therefore, players' preferences do not explain players' strategy choices in VSGs; they rationalize those actions under the condition of strategic interdependence. Players' rationalization and Nash equilibrium together explain stability of interactions.

To illustrate, suppose that Row selects icon in a Peircian game. Besides strategic interdependence and uncertainty, it is impossible to explain Row's selection without going into details of what features of the painting in question appealed her prompting her icon choice. Row should have seen a feature in a, say, Pollock painting, that constitutes the reason rationalizing the choice of icon. Row's selection means that the painting in question constitutes an affirmation of an iconic correspondence with the IR proposition in question. The affirmation reflects the primary reason of the selection of icon or any other choice of sign under strategic interdependence and uncertainty. Game theory supplements the selection of icon such that if Row, for example, thinks that Column would select a sign that is not icon, then Row would rationalize her action generating an outcome of coordination on the basis of, for example, symbol. Reasons rationalize players' strategy choices in a VSG first as there is an appealing value of their actions. The analysis of Davidson (1963) about how reasons rationalize actions in a VSG implies that players' sign choices stem from each player's possession of a pro attitude toward those sign choices and that players believe that those sign choices translate what each player believes in semiotic and game theoretic terms.

---

[9] Italics in the original.

Summing Up

The inexistence of a common language in the Discipline constitutes a major source of conflicting theoretical views about IR. Theoretical terms are sometimes defined and sometimes not. In realist perspective power is understood and defined differently. In constructivism, a key concept of identity does not have unanimously accepted definition. Communication in such a fuzzy and foggy environment is utterly difficult. This book offers the hybrid method of VSGs as a communication tool among IR scholars who understand, and speak the basics of semiotics language. Game theory which neglects sources of players' preferences constitutes a formidable tool to detect convergence of meanings for structural constraints and nonunique dynamic anarchies. Naturally, there are so many other concepts in the discipline open to explorations by the use of VSGs. But the theoretical propositions offered by Waltz and Wendt contain two major concepts open to interpretation. VSGs help to prove the mental coordination on the existence of them. Nash equilibria imply their existence at least in mental terms and in terms of their meanings translated into semiotics.

The discussion notes a parallel between abstract art and structural realism and constructivism. Placements of colors their moves and their meanings help players to reason in their strategic interdependence. The interdependence on the basis of players' preferences over outcomes and actions connects to players' primary and secondary reasons linked to players' awareness of a need for a sound communication constituting mental states of rationalization. The latter helps to solve VSGs by their shaping players' rationalizations. The three mental states are subjective but the players' choices under subjectivities reflect substantiated behavior of players. Players' choices are fueled by these subjectivities.

Future works based on this project can amply widen the scope of VSGs including, for example, dynamic games driven by differential equations, alternative equilibrium concepts and theoretical propositions. The book opens a brand-new interpretive research avenue with an offer of a common language in the Discipline.

© Springer Nature Switzerland AG 2023
S. Ş. Güner, *Art and IR Theory*, Mathematics in Mind,
https://doi.org/10.1007/978-3-031-32342-3

References

Aiello, Giorgia. 2020. "Visual semiotics: Key concepts and new directions." In Luc Pauwels and Dawn Mannay (eds.), *The Sage Handbook of Visual Research Methods*. London: Sage.

Albers, Joseph. 2013. *Interaction of Color: 50th Anniversary Edition*. New Haven and London: Yale University Press.

Antal T., Krapivsky P. L., and S. Redner. 2006. "Social Balance on Networks: The Dynamics of Friendship and Enmity." *Physica D*. 224: 130–136.

Arnheim, Rudolph. 1969. *Visual Thinking*. Berkeley: University of California Press.

Atkin, Albert. 2013. "Peirce's Theory of Signs." *Stanford Encyclopedia of Philosophy*. https://plato.stanford.edu/archives/sum2013/entries/peirce-semiotics.

Audi, Robert. 2015. *The Cambridge Dictionary of Philosophy* (third edition). New York: Cambridge University Press.

Aumann, Robert J. 1987. "Correlated Equilibrium as an Expression of Bayesian Rationality." *Econometrica* 55 (1): 1–18.

Aumann, Robert J. 1985. "What Is Game Theory Trying to Accomplish?" In Kenneth ArRow and S Honkapohja (eds.), *Frontiers of Economics*. Oxford: Basil Blackwell.

Auster, Carol J. 1980. "Balance Theory and Other Extra-Balance Properties: An Application to Fairy Tales." *Psychological Reports* 47: 183–188.

Axelrod, Robert, and Robert O. Keohane.1985. "Achieving Cooperation Under Anarchy". *World Politics* 38 (1): 226–254.

Baele, Stephane J., and Gregorio Bettiza. 2021. "Turning everywhere in IR: on the sociological underpinnings of the field's proliferating turns." *International Theory* 13 (2): 314–340.

Bal, Mieke and Norman Bryson. 1991. "Semiotics and Art History." *The Art Bulletin* 73 (2): 174–208.

Ball, M. S. and Smith, G. W. H. 1992. *Analyzing Visual Data*. London: Sage.

Barthes, Roland. 1972. *Mythologies*. New York: Noonday Press.

Barthes, Roland. 1964. Eléments de sémiology. Seuil: Paris.

Bell, Duncan. 2009. "Writing the World: Disciplinary History and Beyond." *International Affairs (Royal Institute of International Affairs)* 85 (1): 3–22.

Björnerstedt, Jonas and Weibull, Jörgen W. 1994. "Nash Equilibrium and Evolution by Imitation". IUI Working Paper, No. 407, The Research Institute of Industrial Economics (IUI), Stockholm.

Blackburn, Simon. 2016. *Oxford Dictionary of Philosophy*. Oxford: Oxford University Press.

Binmore, Ken. 2007. *Playing for Real: A Text on Game Theory*. Oxford: Oxford University Press.

Bleiker, Roland. 2001. "The Aesthetic Turn in International Political Theory." *Millenium: Journal of International Studies* 30 (3): 509–533.

Blume, S. (1990). "Interdisciplinarity in the Social Sciences." Science Policy Support Group, Concept Paper, no. 10. London.

Boltzmann, Ludwig. 1974. *Theoretical Physics and Philosophical Problems*. Dordrecht: Springer.

Bonanno, Giacomo. 2015. *Game Theory: An open access textbook with 165 solved exercises.*

Brams, Steven J. 1985. *Superpower Games: Applying Game Theory to Superpower Conflict.* New Haven and London: Yale University Press.

Braque. 1997. *Great Modern Masters.* Edited by José Maria Faerna and translated from Spanish by Alberto Curotto. Harry N. Abrams: New York.

Breit, W. 1984. "Galbraith and Friedman: Two Versions of Economic Reality." *Journal of Post Keynesian Economics* 7 (1): 18–29.

Brown, Chris. 2010. *Practical Judgment in International Political Theory: Selected Essays.* London: Routledge.

Butler, Judith. 2007. "Torture and the Ethics of Photography." *Environment and Planning D* 25: 951–66.

Buyssens, Eric. 1967. *La communication et l'articulation linguistique.* Brussels/Paris: Presses Universitaires.

Buzan, Barry and Richard Little. 2001. "Why International Relations has Failed as an Intellectual Project and What to do About it." *Millenium: Journal of International Studies* 30 (1): 19–39.

Callahan, William A. 2015. "The Visual Turn in IR: Documentary Filmmaking as a Critical Method." *Millenium: Journal of International Studies* 43 (3): 891–910.

Caplow, Theodore. 1959. "Further Developments of a Theory of Coalitions in the Triad." *American Journal of Sociology* 66: 488–493.

Cartwright, Dorwin and Frank Harary. 1956. "Structural Balance: A Generalization of Heider's Theory." *Psychological Review* 63 (5): 277–293.

Chalmers, David J. 2022. *Reality+ Virtual Worlds and the Problems of Philosophy.* New York: W. W. Norton and Allen Lane.

Chalmers, David J. 1996. *The Conscious Mind: In Search of a Fundamental Theory.* Oxford: Oxford University Press.

Chandler, Daniel. 2007. *Semiotics: The Basics.* New York: Routledge.

Checkel, Jeffrey. 1998. "The Constructive Turn in International Relations Theory." *World Politics* 50 (2): 324–348.

Christensen, Thomas J. and Jack Snyder. 1997. "Progressive Research on Degenerate Alliances." *American Political Science Review* 91 (4): 919–922.

Churchland, Paul M. 1984. *Matter and Consciousness: A Contemporary Introduction to the Philosophy of Mind.* Cambridge, Massachusetts: MIT Press.

Churchland, Paul M. 1979. *Scientific Realism and the Plasticity of Mind.* Cambridge: Cambridge University Press.

Collingwood, Robin G. 1994. *The Idea of History.* Oxford: Oxford University Press.

Copeland, Dale. 2000. "The Constructivist Challenge to Structural Realism." *International Security* 25 (2): 187–212.

Copi, Irving M. 1961. *Introduction to Logic.* New York: Macmillan.

Cornut, Jérémie. 2017. "The Practice Turn in International Relations Theory." In *Oxford Research Encyclopedia of International Studies.* Edited by Renée Marlin-Bennett. New York: Oxford University Press.

Croce, Benedetto. 1965. *Æsthetic: As Science of Expression and General Linguistic* (Translated by Douglas Ainslie). New York: Noonday Press.

Cross, Charles B. 1991. "Explanation and the Theory of Questions". *Erkenntnis* 34: 237–260.

Danesi, Marcel. 2018. *Of Cigarettes, High Heels, and Other Interesting Things: An Introduction to Semiotics* (Third Edition). New York: Palgrave, Macmillan.

Davidson, Donald. 1980. *Essays on Actions and Events.* Oxford: Clarendon Press.

Davidson, Donald. 1963. "Actions, Reason, and Causes". *The Journal of Philosophy* 60 (23): 685–700.

Davis, Morton D. 1983. *Game Theory: A Nontechnical Introduction.* New York: Basic Books.

De Saussure, Ferdinand. 1996. *Cours de linguistique general*. Payot: Lausanne, Paris.

Descartes, René. 2004. "Minds and Bodies as Distinct Substances." In John Heil (ed.) *Philosophy of Mind: A Guide and Anthology*. Oxford: Oxford University Press, 36–58.

Derry, Gregory N. 1999. *What Science Is and How It Works*. Princeton, New Jersey: Princeton University Press.

Donnelly, Jack. 2019. "Systems, levels, and structural theory: Waltz's theory is *not* a systemic theory (and why that matters for International Relations today)". *European Journal of International Relations* 25 (3): 904–930.

Eco, Umberto. 1976. *A Theory of Semiotics*. Bloomington: Indiana University Press.

Eco, Umberto. 1984. *Semiotics and the Philosophy of Language*. Bloomington: Indiana University Press.

Eichberger, Juergen. 1993. *Game Theory for Economists*. San Diego: Academic Press.

Elkins, James. 2003. "What Does Peirce's Sign System Haver to Say to Art History?" *Culture, Theory and Critique* 44 (1): 5–22.

Farrell, Joseph. 1988. "Communication, Coordination, and Nash Equilibrium." *Economics Letters* 27: 209–214.

Fearon, James and Alexander E. Wendt. 1992. "Rationalism v. Skepticism: A Skeptical View." In Carlsnaes, W., Risse, T. and Simmons, B. A. (eds.) *Handbook of International Relations*. London: Sage.

Feaver, Peter D., Hellman, Gunther, Schweller Randall L., Taliaferro Jeffrey W., Wohlforth, William C., Legro Jeffrey W., and Andrew Moravcsik. 2000. "Correspondence: Brother, Can You Spare a Paradigm? (Or Was Anybody Ever a Realist?)." *International Security* 25 (1): 165–193.

Feldman, Allen. 2005. "On the Actuarial Gaze: From 9/11 to Abu Ghraib." *Cultural Studies* 19(2): 203–226.

Festinger, Leon. 1957. *A Theory of Cognitive Dissonance*. California, Stanford: Stanford University Press.

Fiske, John. 2011. *Introduction to Communication Studies*. London, New York: Routledge.

Floch, Jean-Marie. 1985. *Petites mythologies de l'oeil et de l'esprit: Pour une sémiotique plastique*. Paris, Amsterdam: Hadès-Benjamins.

Fodor, Jerry. 1981. "The Mind-Body Problem." *Scientific American* 144.

Forceville, Charles. 1994. "Pictorial Metaphor in Advertisements". *Metaphor and Symbolic Activity* 9 (1): 1–29.

Friedman, Daniel. 1998. "On Economic Applications of Evolutionary Game Theory," *Journal of Evolutionary Economics* 8 (1): 15–43.

Friedman, Daniel. 1991. "Evolutionary Games in Economics," *Econometrica* 59 (3): 637–666.

Friedman, James W. 1986. *Game Theory with Applications to Economics*. Oxford: Oxford University Press.

Fudenberg, Drew and Jean Tirole. 1991. *Game Theory*. Cambridge, Massachusetts: MIT Press.

Gardner, Roy. 2003. *Games for Business and Economics*. New York: Wiley.

Gauthier, Davis. 1975. "Coordination." *Revue Canadienne de Philosophie* 14 (2): 195–221.

George, Jim. 1995. "Realist `Ethics`, international relations, and post-modernism: Thinking beyond the egoism-anarchy thematic". *Millennium: Journal of International Studies*, 4 (2): 195–223.

George, Jim and David Campbell. 1990. "Patterns of Dissent and the Celebration of Difference: Critical Social Theory and International Relations." *International Studies Quarterly* 34 (3): 269–293.

Gibbons, Robert. 1992. *A Primer in Game Theory*. New York: Harvester, Wheatsheaf.

Giere, Ronald N. 1988. *Explaining Science: A Cognitive Approach*. Chicago and London: The University of Chicago Press.

Gintis, Herbert. 2000. "Classical Versus Evolutionary Game Theory." *Journal of Consciousness Studies* 7 (1–2): 300–304.

Giroux, Henry. 2004. "Education After Abu Ghraib." *Cultural Studies* 18(6): 779–815.

Goddard, Stacie E., Nexon, Daniel H. 2005. "Paradigm Lost? Reassessing Theory of International Politics." *European Journal of International Relations* 11 (1): 9–61.

Godfrey-Smith, Peter. 2003. *Theory and Reality: An Introduction to the Philosophy of Science.* Chicago and London: The University of Chicago Press.

Goldgeier, J. M. and Philip E. Tetlock. 2001. "Psychology and International Relations," *Annual Review of Political Science* 4 (1): 67–92.

Goodman, Nelson. 1968. *Languages of Art.* New York: Bobbs-Merrill.

Goodman, Nelson. 1978. *Ways of Worldmaking.* Indiana: Indianapolis, Hackett.

Greimas, Algirdas J. 1976. *Sémiotique et sciences sociales.* Paris: Seuil.

Groening, Matt. 1987. *The School is Hell: A Cartoon Book by Matt Groening.* Pantheon Books: New York and Random House: Toronto.

Gulick, Edward V. 1955. *Europe's Classical Balance of Power.* New York: W. W. Norton.

Güner, Serdar Ş. 2021. "Wendt Versus Pollock: Towards Visual Semiotics in the Discipline of IR Theory." *Semiotica* 238: 239–251.

Güner, Serdar Ş. 2019. "Waltz Talks Through Rothko: Visual Metaphors in the Discipline of International Relations Theory." *Semiotica* 231: 171–191.

Güner, Serdar Ş. 2017. "An Evolutionary Game Analysis of Balancing and Bandwagoning in Unipolar Systems." *Journal of Game Theory* 6 (2): 21–37.

Güner, Serdar. 2012. "Religion and Preferences: A Decision-theoretic Explanation of Turkey's New Foreign Policy." *Foreign Policy Analysis* 8 (3): 217–230.

Hall, Edward. 1966. *The Hidden Dimension.* Garden City, New York: Doubleday.

Harary, Frank. 1985. "The love-hate structure of Dangerous Corner." *Semiotica* 54 (3–4): 387–393.

Heider, Fritz. 1945. "Attitudes and Cognitive Organizations." *Journal of Psychology* 21(1): 107–112.

Hempel, Carl G. and Oppenheim, Paul. 1948. "Studies in the Logic of Explanation". *Philosophy of Science* 15: 567–579.

Hofbauer, Josef and Sigmund, Karl. 1998. *Evolutionary Games and Population Dynamics.* Cambridge: Cambridge University Press.

Hofbauer, Josef, P. Schuster, and K. Sigmund. 1979. "A note on evolutionary stable strategies and game dynamics." *Journal of Theoretical Biology* 81: 609–12.

Hofstadter, Douglas R. 1985. *Metamagical Themas: Questing for the Essence of Mind and Pattern.* New York: Basic Books.

Hopf, Ted. 1998. "The Promise of Constructivism in International Relations Theory." *International Security* 23 (1): 171–200.

Iversen, Margaret. 2018. "Index and Icon in the Work of Duchamp and Dali". *Avant-garde Studies* 3 (Spring and Summer): 1–14.

Jakobson, Roman. 1960. *Linguistics and Poetics.* In *Style in Language*, ed. Thomas A. Sebeok, 350–377. Cambridge: MIT Press.

Jervis, Robert. 1985. "From Balance to Concert: A Study of International Security Cooperation". *World Politics* 38 (1): 58–79.

Kaplan, Morton. 1966. "The New Great Debate: Traditionalism vs. Science." *World Politics* 19 (1): 1–20.

Kemeny, John G., Snell, Laurie J., Thompson, Gerald L. 1974. *Introduction to Finite Mathematics.* Englewood Cliffs, New Jersey: Prentice Hall.

Kim, Yong-Gwan and Joel Sobel. 1995. "An Evolutionary Approach to Pre-Play Communication." *Econometrica* 63 (5): 1181–1193.

Kirsh, D., Goldin-Meadow, S., Clark, H., & Rogers, Y. 2011. "Interactivity and Thought." *Proceedings of the Annual Meeting of the Cognitive Science Society*, 33. Retrieved from https://escholarship.org/uc/item/0096777s.

Klee, Paul. 1920. "Paul Klee." In Edschmid, K. (ed.), *Tribüne der Kunst und Zeit: Eine Schriftensammlung.* Berlin: Erich Reiss Verlag.

Kreps, David M. 1990. *A Course in Microeconomic Theory.* Princeton: Princeton University Press.

Kurki, M and Wight, C. 2010. "International Relations and Social Science." In Dunne, T., M. Kurki & S. Smith (eds.), *International Relations Theories: Discipline and Diversity*. Oxford: Oxford University Press.

Landau, Ellen G. 1989. *Jackson Pollock*. London: Thames and Hudson.

Langer, Susanne. 1951. *Philosophy in a New Key: A Study in the Symbolism of Reason, Rite, and Art*. Cambridge: Harvard University Press.

Lapid, Yosef. 1989. "The Third Debate: On the Prospects of International Theory in a Post-Positivist Era." *International Studies Quarterly* 33: 235–254.

Legro Jeffrey W. and Moravcsik, Andrew. 1999. "Is Anybody Still a Realist?" *International Security* 24 (2): 5–55.

Lewis, David. 1969. *Convention: A Philosophical Study*. Harvard, Massachusetts: Harvard University Press.

Lisle, Debbie. 2011. "The Surprising Detritus of Leisure: Encountering the Late Photography of War." *Environment and Planning D* 29 (5): 873–90.

Lotman, Yuri M. 1990. *Universe of Mind: A Semiotic Theory of Culture*. Trans. A. Shukman. Bloomington: Indiana University Press.

Luce, R. Duncan, and Howard Raiffa. 1957. *Games and Decisions: Introduction and Critical Survey*. New York: Wiley.

Lyotard, Jean François. 1993. *Toward the Postmodern*. New Jersey: Humanities Press.

Marin, Louis. 2005. *Études sémiologiques: écritures, peintures* (Collection d'esthétique 11). Paris: Braille, Klinksieck.

Martin, Lisa L. 1999. "The Contributions of Rational Choice: A Defense of Pluralism". *International Security* 24 (2): 74–83.

Masters, Roger. 1983. "The Biological Nature of the State". *World Politics* 35 (2): 161–193.

Maynard Smith, John. 1982. *Evolution and Theory of Games*. Cambridge: Cambridge University Press.

McElrath, Richard and Robert Boyd. 2007: *Mathematical Models of Social Evolution: A Guide for the Perplexed*. Chicago: University of Chicago Press.

McGinn, Colin. 2012. *Truth by Analysis: Games, Names, and Philosophy*. Oxford: Oxford University Press.

Mearsheimer, John. 1994. "The False Promise of International Institutions". *International Security* 19 (3): 5–49.

Mercer, Jonathan. 1995. "Anarchy and Identity." *International Organization* 49 (2): 229–252.

Mesquita, Bruce B. and James D MorRow. 1999. "Sorting Through the Wealth of Notions". *International Security* 24(2): 56–73

Milner, Helen. 1991. "The Assumption of Anarchy in IR Theory: A Critique." *Review of International Studies* 17 (1): 67–85.

Mittelberg, Irene. 2006. *Metaphor and Metonymy in Language and Gesture: Discourse Evidence for Multimodal Models of Grammar*. PhD Dissertation, presented to the Faculty of the Graduate School of Cornell University.

Mondrian, Piet. 1951. *Plastic art and pure plastic art, 1937, and other essays, 1941–1943 (The documents of modern art)*. New York: Wittenborn, Schultz.

Mukařovský, Jean. 1977. *Structure, Sign, and Function* (Translated and Edited by John Burbank and Peter Steiner). New Haven and London: Yale University Press.

Musgrave, Alan. 1985. "Realism Versus Constructive Empiricism." In Paul M. Churchland and Clifford A. Hooker, *Images of Science: Essays on Realism and Empiricism, with a Reply from Bas C. Fraassen*. Chicago and London: University of Chicago Press, pp. 197–221.

Myerson, Roger B. 1991. *Game Theory: Analysis of Conflict*. Cambridge, Massachusetts: Harvard University Press.

Nagel, Thomas. 1974. "What Is It Like to Be a Bat?" *The Philosophical Review* 83 (4): 435–450.

Neumann, Iver B. 2002. "Returning Practice to the Linguistic Turn: The Case of Diplomacy." *Millennium: Journal of International Studies*, 31 (3): 627–651.

Nicholson, Michael. 2000. "What's the use of International Relations?" *Review of International Studies* 26: 183–198.

Niou, Emerson M.S. and Ordeshook Peter C. 1999. "Return of the Luddites". *International Security* 24 (2): 84–96.

Nowak, Martin A. 2006. *Evolutionary Dynamics: Exploring the Equations of Life*. Cambridge, MA: Harvard University Press.

Nöth, Winfried. 1990. *Handbook of Semiotics*. Bloomington and Indianapolis: Indiana University Press.

O'Neill, Barry. 2001. *Honor, Symbols, and War*. Ann Arbor: The University of Michigan Press.

O'Neill, Barry. 1987. "A Measure for Crisis Instability with an Application to Space-Based Antimissile Systems." *Journal of Conflict Resolution* 31 (4): 631–672.

Osborne, Martin J., and Ariel Rubinstein. 1994. *A Course in Game Theory*. Cambridge, Massachusetts: MIT Press.

Oye, Kenneth A. 1985. "Explaining Cooperation under Anarchy: Hypotheses and Strategies". *World Politics* 38 (1): 1–24.

Powell, Robert. 1999. "The Modeling Enterprise and Security Studies". *International Security* 24 (2): 97–106.

Peirce, Charles S. 1960a. *Collected Papers of Charles Sanders Peirce (1931–1958), Vol I.: Principles of Philosophy*. Belknap Press of Harvard University Press.

Peirce, Charles S. 1960b. *Elements of Logic, Vol II*. C. Hartshorne and P. Weiss, eds. Belknap Press of Harvard University Press.

Peirce, Charles S. 1960c. "Icon, Index, Symbol." In Charles Hartshorne and Paul Weiss, (eds.) *Collected Papers: Vol. II*. Cambridge, Massachusetts: Belknap Press of Harvard University Press, 161–165.

Peirce, Charles S. 1955. *Philosophical Writings of Peirce: Selected and Edited with an Introduction by Justus Buchler*. New York: Dover Publications.

Peirce, Charles Sanders. 1965. *The Collected Papers of Charles Sanders Peirce*, Vol 2., edited by Charles Hartshorne and Paul Weiss, Cambridge, Massachusetts: Harvard University Press, pp. 247–249.

Perea, Andrés. 2012. *Epistemic Game Theory: Reasoning and Choice*. Cambridge: Cambridge University Press.

Pond, Elisabeth, and Kenneth N. Waltz. 1994. "Correspondence: International Politics, viewed from the ground". *International Security* 19 (1): 195–199.

Rabin, Matthew. 1994. "A Model of Pre-Game Communication." *Journal of Economic Theory* 63: 370–391.

Rapoport, Anatol and Melvin J Guyer. 1966. "A Taxonomy of 2 × 2 Games." *General Systems* 11: 203–214.

Rappaport, Steven. 1995. "Economic Models and Historical Explanation". *Philosophy of the Social Sciences* 25 (4): 421–441.

Renouvin, Pierre. 1945. *Histoire des relations internationales (Tome IV)*. Paris: Hachette.

Riegl, Alois. 1999. *The Group Portraiture of Holland*. Los Angeles, California: Getty Research Institute for the History of Art and the Humanities.

Rose, Gillian. 2001. *Visual Methodologies: An Introduction to the Interpretation of Visual Materials*. London: Sage.

Ross, Lee and Andrew Ward. 1995. "Psychological Barriers to Dispute Resolution." *Advances in Experimental Social Psychology* 27: 255–304.

Rouse, Joseph. 1987. *Knowledge and Power: Towards a political philosophy of science*. Ithaca: Cornell University Press.

Rubinstein, Ariel. 1991. "Comments on the Interpretation of Game Theory." *Econometrica* 59 (4): 909–924.

Ryle, Gilbert. 1949. *The Concept of Mind*. New York: Barnes and Noble.

Sally, David. 2002. "What an Ugly Baby! Risk Dominance, Sympathy, and Coordination of Meaning." *Rationality and Society* 14 (1): 78–108.

Samuelson, Larry. 1997. *Evolutionary Games and Equilibrium Selection*. Cambridge, Massachusetts: MIT Press.

Schapiro, Meyer. 1972\1973. "On Some Problems in the Semiotics of Visual Art: Field and Vehicle in Image-Signs". *Simiolus* 6 (1): 9–19.

Schelling, Thomas. 1980. *The Strategy of Conflict* (Second Edition). Harvard: Harvard University Press.

Schmidt, Brian. 2012. "On the History and Historiography of International Relations." In Carlsnaes, Walter, Risse Thomas and Beth A. Simmons, eds., *Handbook of International Relations*. London, New York: Sage.

Smith, Steve. 2000. "Wendt's World." *Review of International Studies* 26: 151–163.

Schwarz, D. 1997. *Reconfiguring Modernism: Explorations in the Relationship Between Modern Art and Modern Literature*. New York: St. Martin's Press.

Schwartz-Shea, Peregrine. 2006. "Judging Quality: Evaluative Criteria and Epistemic Communities." In Dvora Yanow and Peregrine Schwartz-Shea, eds., *Interpretation and Method: Empirical research Methods and the Interpretive Turn*. New York, London: M. E. Sharpe.

Schweller, Randall L. 1997. "New Realist Research on Alliances: Refining, Not Refuting, Waltz's Balancing Proposition." *American Political Science Review* 91 (4): 927–930.

Scruton, Roger. 1998. *Art and Imagination: A Study in the Philosophy of Mind*. South Bend, Indiana: St. Augustine's Press.

Searle, John R. 1980. "Minds, Brains, and Programs." *Behavioral and Brain Sciences* 3 (3): 417–457.

Sebeok, Thomas A. 2001. *Signs: An Introduction to Semiotics* (Second Edition). Toronto: University of Toronto Press.

Sigmund, Karl. 1993. *Games of Life: Explorations in Ecology Evolution and Behavior*. Oxford: Oxford University Press.

Simmel, Georg. 1955. "Significance of Numbers for Social Life." In Paul A. Hare, Edgar F. Borgatta, and Robert F. Bales (eds.) *Small Groups*. New York: Knopf.

Skyrms, Brian. 2001. "The Stag Hunt". *Proceedings and Addresses of the American Philosophical Association* 75 (2): 31–41.

Slater, Don. 1998. "Analyzing Cultural Objects: Content Analysis and Semiotics." In C. Seale (ed.), Researching Society and Culture. London: Sage, pp. 233–44.

Smith, John M. 1982. *Evolution and Theory of Games*. Cambridge: Cambridge University Press.

Snidal, Duncan. 1985. "The Game *Theory* of International Politics". *World Politics* 38 (1): 25–57.

Sylvester, Christine. 2001. "Art, Abstraction, and International Relations." *Millennium: Journal of International Studies* 30 (3): 535–55.

Thayer, Bradley. 2000. "Bringing in Darwin: Evolutionary Theory, Realism, and International Politics". *International Security* 25 (2): 124–151.

Tickner, Ann J. 2011. "Dealing with Difference: Problems and Possibilities for Dialogue in International Relations." *Millenium: Journal of International Studies* 39 (3): 607–618.

Van Fraassen, Bas C. 1980. *The Scientific Image*. Oxford: Oxford University Press.

Vasquez, John A. 1997. "The Realist Paradigm and Degenerative versus Progressive Research Programs: An appraisal of Neotraditional Research on Waltz's Balancing Proposition." *American Political Science Review* 91: 899–912.

Von Neumann, John, and Morgenstern, Oskar. 1953 (Third Edition). Theory of Games and Economic Behavior. Princeton: Princeton University Press.

Wæver, Ole. 2009. "Waltz's Theory of Theory". *International Relations* 23 (2): 201–222.

Wæver, Ole. 1996. "The Rise and Fall of the Inter-Paradigm Debate." In Dunne, T., Kurki, M., and Smith, S. (eds.). *International Relations Theories: Discipline and Diversity*. Oxford: Oxford University Press, pp. 306–327.

Walt, Stephen M. 1999a. "Rigor or Rigor Mortis: Rational Choice and Security Studies." *International Security* 23 (4): 5–48.

Walt, Stephen M. 1999b. "A Model Disagreement." *International Security* 24 (2): 115–130.

Waltz, Kenneth N. 2000. "Structural Realism After the Cold War". *International Security* 25 (1): 5–41.

Waltz, Kenneth N. 1979. *Theory of International Politics*. Reading, Massachusetts: Addison-Wesley.

Watson, Joel. 2008. *Strategy: An Introduction to Game Theory*. New York: Norton.

Wendt, Alexander. 2000. "On the Via Media: a response to the critics." *Review of International Studies* 26 (1): 165–180.

Wendt, Alexander. 1999. *Social Theory of International Politics*. Cambridge: Cambridge University Press.

Wendt, Alexander. 1998. "On constitution and causation in International Relations." *Review of International Studies* 24: 101–117.

Wendt, Aexander. 1995. "Constructing International Politics." *International Security* 20 (1): 71–81.

Wendt, Alexander. 1992. "Anarchy is what states make of it: the social construction of power politics." *International Organization* 46 (2): 391–425.

Wildavsky, Aaron. 1987. "Choosing Preferences by Constructing Institutions: A Cultural Theory of Preference Formation." *American Political Science Review* 81 (1): 3–21.

Williams, Michael C. 2018. "International Relations in the Age of Image." *International Studies Quarterly* 62: 880–891.

Yanow, Dvora. 2006a. "Thinking Interpretively: Philosophical Presuppositions and the Human Sciences." In Yanow, D. and Schwartz-Shea, P. (eds.). *Interpretation and Method: Empirical Research Methods and the Interpretive Turn*. New York, London: M. E. Sharpe, pp. 5–26.

Yanow, Dvora. 2006b. "How Built Spaces Mean: A Semiotics of Space." In Yanow, D. and Schwartz-Shea, P. (eds.). *Interpretation and Method: Empirical Research Methods and the Interpretive Turn*. New York, London: M. E. Sharpe, pp. 349–366.

Zagare, Frank C. 1999. "All Mortis, No Rigor." *International Security* 24 (2): 107–114.

Zajonc, R. B. (1980). "Feeling and Thinking: Preferences Need No Inferences." *American Psychologist* 35: 151–175.

Zinnes, Dina A. 1976. *Contemporary Research in International Relations: A Perspective and a Critical Appraisal*. New York and London: The Free Press.

Printed in the United States
by Baker & Taylor Publisher Services